Mapping Montessori Materials for AI Competency Development

Kate Broughton

KATE BROUGHTON

Kate Broughton Ltd.

ISBN: 978-1-966665-00-7

Independently Published by Kate Broughton Ltd.

Book Cover Design by – Original Design by Book Brush *(edits by A. Broughton)*

For permissions, inquiries, or additional information, please contact:
Email: hello@katebroughton.com
Website: www.katebroughton.com

Published in the Unites States of America

Access Your Resources

Visit **www.katebroughton.com/mappingresources** and enter the password below to access downloadable materials, templates, and tools to accompany this book.

Password: MmMfAiCd26

Version: Living document — Updated July 2026

Contents

Chapter 1

Disclaimer

THIS IS A WORKING DOCUMENT

This publication was developed between February 2023 and 2025 as part of an independent research initiative undertaken during the American Montessori Society (AMS) Innovation Fellowship Think Tank (IFTT). While the work reflects insights developed during that period, the views, interpretations, and conclusions expressed herein are solely those of the author and do not represent the official position, endorsement, or policy of the American Montessori Society or any affiliated organization.

The content provided in this book is intended for informational and educational purposes only. It does not constitute professional advice, and should not be interpreted as prescriptive guidance for any individual, school, or institution. The ideas presented are exploratory in nature and are offered to stimulate reflection, conversation, and adaptation within each unique educational context.

No part of this publication should be construed as a directive or legally binding instruction. Each reader is responsible for independently evaluating the relevance, applicability, and safety of the practices discussed, in accordance with local policies, legal regulations, and the needs of their school community. The decision to engage with or implement any of the material is entirely voluntary and at the reader's discretion.

This work also demonstrates the use of generative artificial intelligence (ChatGPT 4.0 & 5.0) as a collaborative research and drafting assistant. While AI was used to support some formatting, synthesis, and idea development, all content has been reviewed and shaped by the author. This collaborative process is intended to highlight the evolving potential of human-AI partnerships in ethical, transparent, and human-centered educational work.

Chapter 2
A Thread of Humility

Before you read further, I want to offer a simple but heartfelt apology for any mistakes, misspellings, mispronunciations, misunderstandings, or omissions that may appear in these pages. This work, like all human creations, is imperfect. It carries the weight of best intentions, yet it is bound by the limitations of human effort, perspective, and cultural positioning. And so, I ask that you receive this book not as a flawless document, but as an offering, from my heart to yours.

I do not stand alone in this acceptance of imperfection. I stand on the shoulders of cultures and traditions that have long recognized that flawlessness is not the goal. It is often the mistakes, the missteps, and the small irregularities that give something its soul, its protection, and its meaning.

Across history, artisans have deliberately woven humility into their work:

- In **Persian and Islamic rug making**, weavers sometimes introduce a deliberate flaw, often referred to as a "Persian flaw," as a spiritual reminder that only the Divine can create something perfect. To claim otherwise would be an act of arrogance.

- The **Navajo (Diné)** people often include a “spirit line” or “weaver’s path” in their textiles, an intentional line leading out of the design. This allows the spirit of the weaver to remain free, symbolizing a connection to a greater spiritual flow and the sacredness of balance.

- In **Japanese culture**, the concept of *Wabi Sabi* embraces the imperfect, the transient, and the incomplete. Cracks, asymmetry, and signs of age are not flaws, they are the very marks of authenticity and beauty.

These traditions remind us that perfection is not always noble. Sometimes, it is the flaw that protects the spirit, the deviation that preserves our humanity.

I recognize that my understanding of these practices is shaped by an outsider's lens. I offer these references with deep respect, knowing they reflect living traditions with layers far beyond my brief descriptions. If I have misunderstood or misrepresented anything, I welcome correction, dialogue, and deeper insight.

This book carries its own spirit lines and flaws. Despite careful review, there will be errors, of language, of understanding, of cultural context. I may not always say things the right way. I ask for your patience, and I thank you for your grace.

But like the artisans before me, I offer this work as a living conversation. These words are threads, woven into something larger than myself. They will take new shape in your hands, shaped by your insights, your community, your children.

And so, if you find an error, I invite you to hold it not as a flaw, but as a thread of humility, a reminder that we are all learning, growing, and weaving this future together.

In Montessori, the adult learns to observe without ego, to teach without imposing, and to follow the child with reverence. In a world increasingly shaped by artificial intelligence, holding space for human imperfection may be one of the most powerful tools of discernment we already possess.

May this work inspire curiosity, conversation, and collaboration. And may it be received, as it is given, from my humble heart to yours. I am proud to introduce this mapping work to you and look forward to our journey together.

Chapter 3

Introduction

The rapid advancements in artificial intelligence and emerging technologies have reshaped society at a pace few could have anticipated. In just five years, generative Artificial Intelligence (AI), automation, robotics, and machine learning have disrupted longstanding assumptions about work, learning, and human development. We now find ourselves asking not whether education must evolve, but how we can ensure that the environments entrusted with children's growth remain deeply humane, future-conscious, and anchored in principles that honor human potential.

This book enters that question through the Montessori approach, while extending far beyond Montessori circles. It is written for every educator, school leader, policymaker, parent, and student, anyone who calls the classroom home or cares about the future of childhood and learning. Montessori offers not a nostalgic lens but a living, dynamic, scientifically grounded pedagogy capable of meeting the philosophical, cognitive, and ethical demands of an AI-augmented age.

This work began in February 2023 through the newly formed American Montessori Society (AMS) Innovation Fellowship Think Tank (IFTT)(2023–2025). A first draft of the mapping was presented at The Montessori Event in Florida in 2024, then expanded and refined for The Montessori Event in Denver in 2025, as a central part of the AMS 65th Anniversary Celebrations within the exhibit Innovation Labs: The Future Montessori Classroom, which shaped partnerships with educators seeking clarity and confidence in this era of deep technological transition. Yet for me, the journey did not begin with AI; it began forty years ago, in classrooms in the United Kingdom and later in the United States, observing children, founding and leading a Montessori school, guiding teacher education programs, and serving as faculty for the Gulf Coast Montessori Teacher Education Center (Elementary) and the Montessori Education Center of the Rockies (Administrator).

In recent years, my work extended beyond the Montessori ecosystem into AI leadership and strategy, including serving as an advisor in the Strategic AI Program with Z School alongside higher education and industry innovators. These experiences deepened a simple, enduring truth: Montessori already cultivates the cognitive and ethical architecture children need in the age of intelligent machines. The work now is not to reinvent the method, but to articulate its relevance with precision and courage, and to accompany educators and families as they translate timeless human development principles into contemporary practice.

Because clarity brings grounding, an early chapter in this book introduces a Montessori–AI Triage: a simple hierarchy of what matters most as schools navigate digital tools and AI adoption. It centers children's developmental needs, protects concentration and human connection before technology, and reminds us that no tool should outrun the prepared adult or the prepared environment, whether physical, digital, or internal.

From there, the book unfolds through a living framework that explores a Montessori Triage model, Dr. Montessori's life and insights, the definition and history of AI, the Montessori Prepared Environment, the idea of Cosmic Task, the Montessori materials, the Great Lessons, and the developmental planes, showing why this mapping matters and why now. It aligns these foundations with global AI competency frameworks, including those from UNESCO, the World Economic Forum, and McKinsey, positioning Montessori squarely within the global education conversation rather than adjacent to it.

The chapters weave together literature review, historical grounding, methodology, reflective inquiry, material mapping across ages birth to eighteen, extensions and applications, competency rubrics, classroom posters, a glossary of Montessori concepts connected to AI, and two companion glossaries of Montessori, AI, and educational terminology. Combined they offer language, tools, and perspectives to support educators as researchers, families as partners, and young people as ethical and capable makers of the world they will inherit.

Montessori is a research-based pedagogy rooted in empirical observation, not tradition for tradition's sake. It cultivates executive functioning, imaginative capacity, and moral agency alongside precision, rigor, and deep respect for the child's inner life. These are the same qualities increasingly recognized as essential in an AI-shaped society. Montessori does not prepare children to compete with machines; it prepares them to lead, to create, to discern, to collaborate, and to remain fully human in a technological world.

What follows is not a conclusion, but an invitation, a beginning draft of a shared language and shared responsibility. Montessori educators, parents, school leaders, policymakers, researchers, and students themselves each have a role in shaping a future where technology amplifies human dignity and intelligence rather than diminishing it. May this work help illuminate the gold embedded in the writings and vision of Dr. Maria Montessori, and may it serve as a compass as we navigate the profound and urgent challenges and possibilities of our time.

Chapter 4

A Living Framework

With these principles in mind, we move into the applied dimensions of this model, not as a community seeking certainty, but as one committed to research, collaboration, and responsiveness to the whole child. What follows belongs to all of us, to be tested, refined, questioned, and strengthened through real work with real children.

This book is not a rigid prescription, but a flexible framework. It invites exploration at the intersection of Montessori education and emerging technologies, encouraging thoughtful experimentation and deep inquiry into what it means to guide children in this technological age.

It offers a series of sample lessons organized by age group and competency, not as mandates but as possibilities. Each lesson follows a structure Montessori educators will recognize: clear prerequisites, materials, presentation steps, language, and points of interest, alongside control of error, variations, and real-world connections. These examples are meant to feel familiar and practical while honoring the evolving needs of today's learners and the increasing complexity of the tools that shape their world. They are an invitation to adapt, iterate, and create, not to replicate, so that Montessori environments remain living laboratories of inquiry, agency, and ethical engagement in a changing era.

You might choose to engage with this book in a variety of ways:

- As a spark for school-wide conversations around digital

literacy and ethical technology use

- Through focused staff meetings or professional development
- By age level (Infant–Toddler, Early Childhood, Elementary, or Adolescence)
- As part of parent education programs

As you read, you will find that many of these ideas are already alive in your Montessori environment; that is by design. This work is not about reinventing Montessori, it is about revealing how completely the method already prepares children for this moment in history.

Breaking the Sequence: A Montessori Lens on Learning Paths

Montessori educators use a structured lesson format that supports clarity, sequencing, and guided exploration. Across trainings and handbooks, you will find familiar elements:

- A Title – defining the core idea
- A Direct Aim – the immediate learning goal
- An Indirect Aim – subtler developmental benefits
- Prerequisites – prior knowledge or readiness
- Materials – the hands-on tools for discovery
- Presentation – a step-by-step approach
- Variations & Extensions – to deepen or broaden the experience
- What Comes Next – natural follow-up lessons

- Control of Error – built-in feedback for independence

This structure has supported generations of guides. It helps ensure children build knowledge progressively, with opportunities for revisiting earlier concepts and branching forward into new ones.

But what happens when a child does not follow this path?

When the Child Leads, Not the Lesson

Some children barely touch the materials, yet fully grasp the concept, observing from a distance and absorbing everything. I have taught students who consistently scored in the 99th percentile on standardized tests without completing the age related Montessori sequence of lessons.

How do we account for that?

- For seasoned guides, this might affirm something you have always sensed: children do not all need the same pathway. For newer educators, it might raise a challenge: if the child moves away from the material, are we still doing it "right"?

- For public school educators, it opens the door to more fluid learning models that challenge the notion that learning must be linear.

- For researchers, it invites questions about engagement, attention, and the nature of understanding itself.

- And for EdTech developers, it presents an urgent opportunity: how can we build AI systems that honor the flexibility and responsiveness Montessori has always offered, rather than asking the learner to conform to the machine?

Why This Book Looks Different

Because of these realities, I have chosen to organize this book differently. The lessons in this book are structured not by traditional academic subjects, but by competencies, skills children need in a world shaped by AI. Within each competency, lessons are offered in a structured Montessori format, but are meant to be dynamic and responsive.

They are not prescriptions, they are provocations. They are here to be adapted, rewritten, rejected, or rediscovered depending on your context and your students.

- If you are new to Montessori, this book will show you how we structure lessons, and why.
- If you are a seasoned guide, it may remind you of the freedom within the form.
- If you work in a more traditional educational setting, I invite you to see where flexibility might live within your structures.
- If you are a researcher, explore how learning happens when we loosen the grip of sequential expectations.
- If you are designing AI for education, ask yourself: what would it take to build a system that follows the child?

Take This, Make It Your Own

- This book is a living framework, intended to grow with you.
- If you need structured lessons, use them.
- If you need inspiration, read the reflections between the lines.
- If you are looking for certainty, you will not find it here.

- What you will find is a way of seeing.

Because real learning does not come from following rules. It comes from observing what is real, responding with presence, and trusting the unfolding of human development in all its beautiful, unpredictable forms. If we want to learn from a master of unpredictable forms, we may look no further than Dr. Maria Montessori herself.

Chapter 5
Montessori Triage Framework

"Education is a natural process carried out by the child and is not acquired by listening to words."
— Dr. Maria Montessori

Why Triage? Why Now?

While working with IFTT researcher Michelle Velho, we began using the phrase "Montessori Triage" to describe a sequence of decision-making and care that places the child's developmental needs at the center of educational practice. The term "triage" is traditionally used in healthcare to prioritize care based on urgency and foundational needs. In our context, it became a powerful metaphor for reclaiming Montessori's original order of operations: *observe the child, prepare the environment, support character development, and then introduce academic content.*

Rather than reacting to academic pressure, testing mandates, or the acceleration of AI in schools, Montessori Triage reminds us to pause and look at what is most essential first: the human being in front of us. The Montessori method was never meant to begin with measurement or curriculum, it begins with the child. In this framework, triage does not imply emergency; it implies clarity, sequencing, and respect for natural development.

We offer this Montessori Triage as both a protective posture and a practical guide: a way to uphold dignity, agency, and developmental integrity in Montessori environments, especially in a time of rapid technological change.

Montessori educators today are navigating complex challenges, including the growing presence of technology and artificial intelligence in the classroom. This triage is not a new framework, it is a rearticulation of Montessori's original developmental sequence, offered with renewed clarity and purpose. It provides a structured lens for making decisions that protect human development in a fast-paced, technology-driven world.

The Montessori Triage restores focus to the natural order of learning:

- Observation
- Prepared Environment
- Character Formation
- Academic Mastery

When used consistently, this order helps schools resist the pressure to accelerate outcomes at the expense of development. It protects what Montessori called **"the secret of childhood."**

Level One: Observation & the Prepared Environment

Focus: Developmental Readiness

“The first essential for the child’s development is concentration.” — Dr. Maria Montessori

Guide Commitments:

- I observe before intervening.
- I design the environment before designing lessons.
- I protect focus and minimize interruptions.
- I trust the child’s natural development.

Core Practices:

- Daily observation logs.
- Child-led activity selection.
- Uninterrupted work periods.
- Mixed-age classrooms to support peer learning.

Questions to Ask:

- What is this child showing me through action or inaction?
- Is the environment supporting concentration and independence?
- Have I prepared the space before preparing the content?

Translation for Practice:

- Developmentally appropriate pacing

- Constructivist learning models (see chapter 61 for research paradigms)
- Environments as active teaching tools

Level Two: Process & Character Formation

Focus: Social and Emotional Development

"We must help the child act for himself, will for himself, think for himself." — Dr. Maria Montessori

Guide Commitments:

- I support independence and intrinsic motivation.
- I model calm, respectful behavior.
- I allow for productive struggle as part of learning.

Core Practices:

- Choice-based activities.
- Grace and courtesy lessons.
- Conflict resolution and peer mediation.
- Repetition and follow-through to build resilience.

Questions to Ask:

- Is this child demonstrating autonomy or compliance?
- Are social-emotional skills being practiced naturally?
- Is this environment emotionally safe?

Translation for Practice:

- Social-emotional learning integration
- Mastery-based progression
- Respect for developmental differences

Level Three: Academic Expression and Mastery

Focus: Cognitive Growth and Skill Application

“The greatest sign of success for a teacher... is to be able to say, ‘The children are now working as if I did not exist.’” — Dr. Maria Montessori

Guide Commitments:

- I assess to support growth, not to control it.
- I value understanding and application over rote recall.
- I support the pace of mastery, not deadlines.

Core Practices:

- Student-led presentations
- Portfolios and observation-based assessment
- Progress tracking based on readiness, not age

Questions to Ask:

- Has this child internalized the concept?
- Are we supporting deep understanding or surface compliance?
- Are assessments aligned with Montessori developmental principles?

Translation for Practice:

- Authentic assessment methods
- Competency-based learning

- Documentation that reflects the whole child

Leadership Application: Triage for Montessori School Leaders

To lead a Montessori school in the age of AI and rapid change is to protect the conditions that allow for deep human learning.

Leadership Focus:

- Guard developmental time from being compressed
- Empower guides to observe before intervening
- Align school policy with Montessori philosophy
- Use technology to support, not replace, relationships

Weekly Leadership Reflection:

- Are classrooms structured for independent engagement?
- Are guides using observation to inform instruction?
- Are we protecting pace, agency, and dignity in policy decisions?
- Is technology aligned with Montessori values and purpose?

Guiding Principle:

Support development first. Measure progress second. Never sacrifice pace for performance.

A Montessori Promise for Families

(Parent-Facing Language)

At our school, learning begins not with tests or pressure, but with trust in your child's unique path.

What We Prioritize:

- **Observation:** We spend time learning who your child is before deciding what they need.
- **Environment:** We prepare spaces that promote focus, movement, and meaningful activity.
- **Relationships:** We build trust and respect between children, educators, and families.
- **Inner Motivation:** We support curiosity, not compliance.

What This Means for Your Child:

- Progresses at a developmentally appropriate pace
- Builds independence, focus, and resilience
- Develops real-world problem-solving and collaboration
- Experiences meaningful, hands-on academic learning

Our Commitment:

- We observe before acting.
- We design learning environments with intention.
- We support effort, not just outcomes.
- We partner with families, not override them.

About Technology:

Technology helps our staff with planning and communication, but your child's learning is grounded in real-world interaction, not screen-based delivery. We believe people, not devices, raise thinkers.

Bridge to Competency Mapping

With this triage as a foundation, the next section of this book will explore how Montessori materials naturally support the competencies required for today's world, especially in the age of AI. These include pattern recognition, decision-making, data literacy, and ethical reasoning. We begin not with tools or outcomes, but with the child, and from that foundation, we build a future-ready education.

Chapter 6

Dr. Maria Montessori: Her Life & Legacy

With the triage established and the pathway illuminated, we return to the source, to the incredible woman who first taught us to see the child before the system, the human before the outcome. Dr. Maria Montessori's revolutionary contributions to education continue to shape classrooms, curricula, and philosophy around the world. This chapter offers a concise yet respectful overview of her life, the development of the Montessori method, and the enduring relevance of her work. In revisiting her story and scientific foundations, we deepen our understanding of the living legacy that guides our practice today.

Born in 1870 in Chiaravalle, Italy, Maria Montessori was a trailblazer from the outset. As a teenager, she enrolled in a technical school for boys in Rome to study engineering, an unconventional path that met with strong resistance from her parents, particularly her father, who hoped she would pursue a more traditional role for women. Montessori persisted, later shifting her focus to medicine.

Dr. Maria Montessori was not the first woman physician in Italy, but she was the first woman to graduate in medicine from the University of Rome in 1896. Her groundbreaking role lies not in being the first, but in transforming medicine and education through her scientific approach to child development. Her early medical career brought her to the Orthophrenic School, a clinic and educational institution for children then referred to as "feeble-minded." There, she applied scientific observation to her work with children with cognitive disabilities and developmental delays. She discovered that when given structured materials and the freedom to engage purposefully, these children made remarkable progress, often outperforming their peers in traditional classrooms.

These experiences ignited Montessori's conviction that all children, regardless of ability, thrive when offered purposeful activity, freedom within limits, and a prepared environment. Her scientific training and keen observational skills led her to conclude that education should not be imposed but discovered, rooted in the child's natural developmental drive (Montessori, 1912/1964).

During her university years, Montessori also attended lectures in anthropology and psychology, particularly those of Giuseppe Sergi, an influential anthropologist who shaped her view that education must be grounded in human biological and cultural development. Montessori's early publication *Pedagogical Anthropology* (1913) framed education as a method of social reform based on scientific inquiry. She described educators as observers and experimenters, and children as the builders of civilization.

In 1907, Dr. Montessori opened her first *Casa dei Bambini* (Children's House) in the working-class district of San Lorenzo, Rome. Here, she applied her ideas with typically developing children, crafting what she called the *prepared environment,* a learning space designed to promote independence, concentration, and self-discipline. The success of the *Casa dei Bambini* drew international attention and marked the formal beginning of the Montessori method.

Dr. Montessori's educational approach was based on deep respect for the child's innate abilities and the scientific study of development. As she explained, "Education is a natural process carried out by the child and is not acquired by listening to words but by experiences in the environment" (Montessori, 1949/1995).

Her method is characterized by several core principles:

- **Respect for the Child** – Valuing each child as a unique individual capable of self-construction.
- **The Absorbent Mind** – Recognizing the child's natural ability to unconsciously absorb knowledge, particularly in the early years (Montessori, 1949/1995).
- **Sensitive Periods** – Identifying optimal developmental windows in which children are particularly receptive to acquiring specific skills (Montessori, 1936/1996).
- **Prepared Environment** – Structuring the environment with freedom, order, and beauty to encourage exploration and purposeful activity.
- **Autoeducation (Self-Directed Learning)** – Understanding that learning arises spontaneously from within the child when provided with the right materials and support.

Dr. Montessori emphasized that learning is not a linear transmission of knowledge, but a process of *psychic unfolding,* the development of the whole human being. Her holistic vision embraced physical, emotional, intellectual, and spiritual dimensions of human potential. She considered education a means for peace, stating, "Establishing lasting peace is the work of education; all politics can do is keep us out of war" (Montessori,1949/2007).

In 1915, Dr. Montessori presented a fully functioning *Glass Classroom* at the Panama-Pacific International Exposition in San Francisco. Enclosed in glass walls, the live demonstration allowed thousands of spectators to observe children engaged in independent, purposeful work, providing compelling evidence of the Montessori method's effectiveness and drawing international acclaim.

The Montessori movement spread rapidly across Europe, the

United States, and Asia. To protect the integrity of her method, Dr. Montessori founded the Association Montessori Internationale (AMI) in 1929. Later, in 1960, American educator Nancy McCormick Rambusch established the American Montessori Society (AMS) to adapt Montessori's principles to the cultural and regulatory context of the United States. Today, AMI and AMS remain the two most recognized organizations supporting Montessori educator preparation and school standards.

Dr. Montessori's influence continues to grow, yet her method has also faced critique. Some educators argue that its emphasis on child-led learning lacks sufficient structure for certain learners. Others point to the cost of materials and teacher training as barriers to accessibility. Nonetheless, research increasingly supports Montessori's efficacy. Studies have shown that Montessori education fosters academic achievement, creativity, executive function, and socio-emotional development, especially in underserved communities (Lillard, 2005; Marshall, 2017).

In recent decades, neuroscience and developmental psychology have validated many of Dr. Montessori's early insights. Concepts such as neuroplasticity, sensitive periods, and the role of movement in learning affirm her belief that early experiences shape the architecture of the brain. In this light, Dr. Montessori was not merely an educator, but a scientist, philosopher, and humanitarian whose work remains profoundly relevant in the 21st century.

In March 2025, the AMS 65th Anniversary Celebrations, at The Montessori Event in Denver (Colorado, US), hosted *Innovation Labs: The Future Montessori Classroom as a* tribute to Dr. Montessori's profound legacy. The exhibit served as a contemporary response to Montessori's original vision, demonstrating the continued vitality and evolution of her pedagogy in the 21st century.

As Paula Polk Lillard aptly stated, "Montessori education is an aid to life, helping children reach their full potential." Dr. Montessori herself offered a visionary call to educators and leaders:

“The child is both a hope and a promise for mankind.”

Her work invites us to see the child not only as a learner, but as a builder of peace, society, and the future. As we look to the future, the question arises: how can Montessori’s time-tested pedagogy respond to the opportunities and challenges presented by emerging technologies and artificial intelligence?

Chapter 7

Definition & Short History of Artificial Intelligence

Artificial Intelligence (AI) has intrigued humanity for centuries, transitioning from ancient myths of intelligent automata to today's advanced machine learning technologies (Crevier, 1993; McCorduck, 2004). Understanding AI's historical progression assists in responsibly navigating its future applications and innovations.

The philosophical foundations of AI emerged from inquiries by thinkers such as René Descartes and Gottfried Wilhelm Leibniz, who debated the possibility of machines replicating human thought processes. Descartes' philosophical assertion, "I think, therefore I am," significantly influenced early discussions about intelligence (Jones, 2015).

AI formally originated during the Dartmouth Conference of 1956, spearheaded by John McCarthy, Marvin Minsky, Nathaniel Rochester, and Claude Shannon, who collectively proposed the hypothesis that machine intelligence could simulate human intelligence, marking AI's inception as an academic discipline (McCarthy et al., 1956).

Early AI systems like ELIZA, a pioneering chatbot developed by Joseph Weizenbaum, showcased initial natural language processing capabilities. Expert systems such as MYCIN, designed to assist medical diagnostics, illustrated AI's potential utility in specialized fields. Nevertheless, early limitations precipitated "AI winters," characterized by diminished funding and waning interest during the 1970s and 1980s (Crevier, 1993; McCorduck, 2004).

AI research revitalized in the 1990s, propelled by advancements in computational power and algorithmic refinement. A significant breakthrough occurred in 1997 when IBM's Deep Blue defeated world chess champion Garry Kasparov, underscoring AI's capabilities in sophisticated decision-making contexts (Campbell, Hoane, & Hsu, 2002; Kasparov, 2017).

Subsequent developments in machine learning and deep learning substantially advanced AI. Milestones such as Google's AlphaGo, which triumphed over human champions in the strategic game of Go, and Transformer-based models have revolutionized AI's applications across domains such as language processing, healthcare, and automation (LeCun, Bengio, & Hinton, 2015).

Despite AI's extensive benefits, significant challenges remain, including algorithmic biases, data privacy issues, potential job displacement, and profound ethical concerns. Researchers persistently investigate explainable AI methodologies, model fairness, and the ambitious goal of achieving Artificial General Intelligence (AGI), a system characterized by versatile, human-like reasoning across various domains (Adadi & Berrada, 2018; Barocas & Selbst, 2016; Goertzel, 2014).

The evolution of AI mirrors humanity's enduring aspiration to comprehend and emulate intelligence. From philosophical contemplation to present-day advancements, AI continually reshapes various sectors, communication methods, and everyday life experiences. As society ventures deeper into an AI-driven epoch, prioritizing responsible development and ethical oversight remains paramount. Within this context, Montessori education offers a powerful counterbalance, rooted in observation, hands-on exploration, and human development over time. By examining Montessori materials through the lens of emerging technologies, we can begin to map how these tools already cultivate many of the core competencies needed in an age shaped by AI.

Chapter 8

The Prepared Environment

The prepared environment is one of Montessori's most radical contributions to education and remains the foundation of her method. Rather than centering the teacher, Montessori centered the child's natural developmental powers and built an environment to support them. She described this environment as a place "where the child can live his own life" and construct himself freely through purposeful activity (Montessori, 1912/2007; Montessori, 1948/1995). In this vision, the room becomes a carefully orchestrated ecosystem in which every detail serves the child's independence, concentration, and dignity.

Historically, Montessori's inspiration grew from her medical training and early work with children in Rome's psychiatric clinic, the Orthophrenic School (*Scuola Ortofrenica*). There she observed that children flourished when given sensory materials and freedom of movement rather than restraint and correction (Kramer, 1976; Standing, 1957). When she established the first Casa dei Bambini on January 6, 1907, on Via dei Marsi in the San Lorenzo district of Rome, she applied this insight to typical children, creating child-sized tables and chairs, placing materials on low shelves, and offering real objects to support meaningful activity (Montessori, 1912/2007; Kramer, 1988). She hired local carpenters to build furniture that children could move themselves and commissioned skilled craftspeople to shape wood, metal, and textiles to her precise design (Standing, 1957). Where no suitable tools or objects existed, she invented them, often observing a child's spontaneous interest and then creating or refining a material directly in response. The first cylinders, tactile letters, and early sensorial apparatus emerged not from abstract theory, but as material answers to the child's demonstrated needs (Montessori, 1946/2012). The physical environment therefore became the pedagogue, designed to respond to the "inner teacher" within each child (Montessori, 1946/2012).

The prepared environment is characterized by beauty, simplicity, and order. Montessori observed that young children are deeply sensitive to order in their surroundings, and that an uncluttered, harmonious space supports calm, orientation, and concentration (Montessori, 1948/1995; Montessori, 1936/1973). Natural light, soft tones, and large windows connect the child to the world beyond the classroom walls, dissolving the artificial boundary between indoors and outdoors. Montessori wrote that children "must feel themselves in contact with the universe" and that nature offers "a great cosmic laboratory for education" (Montessori, 1967). Plants fill the space, often brought from home by the children, who take pride in caring for them as part of the daily life of the classroom. Shelves are curated rather than crowded. Materials are complete and mathematically sequenced, displayed with space around them to invite careful, intentional work (Montessori, 1909/2007; Gettman, 1987). Light is natural whenever possible, objects are clean and well-maintained, and natural materials predominate, offering sensorial clarity and aesthetic grounding (McTamaney, 2009; Helfrich, 2018). The room speaks softly and invites the child into purposeful engagement.

Importantly, the indoors extend naturally outside. Montessori environments include gardens for flowers, herbs, fruit, and vegetables; quiet pathways for walking meditation; and open spaces for cooperative games and natural loose-parts play. As early as the 1910s, Montessori's work in Scandinavia with educational patron Gösta Mittag-Leffler helped model garden-rich school grounds where children cultivated plants, gathered natural specimens, and worked outdoors as a natural extension of classroom life (Standing, 1957; Montessori, 1948/1995). Montessori insisted that "there is no description, no image in any book that is capable of replacing the sight of real trees" and that time in nature awakens "the soul of the child" (Montessori, 1948/1995). Children tend soil, water seedlings, harvest food, and observe seasonal change, integrating biology, ecology, practical life, responsibility, and reverence for nature. Silence exercises and walking meditation cultivate inward stillness, sensory attunement, grace, and presence of mind, practices Montessori integrated to nurture the child's inner life as much as their intellectual one.

At the heart of this environment are the Montessori materials, each scientifically designed to isolate a concept, contain a built-in control of error, and support the child's development through movement and repetition (Montessori, 1949/2007; Duffy, 2007). Practical life activities introduce real tools for real work: pouring water, polishing wood, washing cloth, tending plants, preparing food. These are not imitations but authentic tasks that cultivate motor control, executive functioning, care of the self and environment, and a quiet inner discipline (Stephenson, 2010; McTamaney, 2009). In keeping with this commitment to reality, even the youngest children typically use real glass cups and pitchers in Montessori environments. This practice has been present since Montessori's earliest classrooms and continues in toddler and preschool settings today (Hainstock, 1997; Stephenson, 2010). The purpose is developmental, not aesthetic; glass invites precision, responsibility, and intentional movement, and its presence communicates trust in the child's capability. When glass breaks, the moment becomes a calm and dignified opportunity for learning and care for one's environment, reinforcing both practical skill and inner composure (Stephenson, 2010; Gerber, 2002). Through the experience of handling fragile, beautiful materials, children discover that their actions have meaning and that they are worthy of real tools.

An example of early childhood work: Leaf Polishing: One Lesson, Many Capacities

A small plant is carried to the table with two hands. The child arranges a mat, a bowl of water, cotton pads, and a soft drying cloth. With careful movements, the child dips, presses, wipes, and dries each leaf, removing dust so the plant may breathe and receive light. In this single act the child practices: fine-motor coordination, sequencing, concentration, care of the environment, botany knowledge, observation, classification, patience, responsibility, aesthetic appreciation, sensory refinement, executive functioning, emotional regulation, and ecological empathy. One humble gesture, a constellation of human capacities.

Sensorial materials refine perception and classify experience; mathematical materials move from concrete to abstract; language, geography, biology, and art are offered through concrete, beautiful, precise objects that awaken the intellect through the hand (Montessori, 1917/1992; Duffy & Duffy, 2002; Richardson, 1953). The physical world becomes a key to the child's mind.

Crucially, the prepared environment also includes open floor space for rugs and large work, accessible outdoor areas, child-height sinks and shelves, and carefully chosen grace and courtesy tools such as quiet movement pathways and defined personal work spaces (Montessori, 1936/1994; Montessori, 1966/1994). Montessori insisted that freedom must be real, not symbolic: when a child can reach a pitcher, carry a broom, open a cupboard, and choose meaningful work without asking permission, independence shifts from an ideal to a lived daily experience (Montessori, 1949/1994). The adult's role, therefore, is not to command but to prepare, observe, and maintain this environment with reverence, knowing that the child's self-formation unfolds through it (Standing, 1957; Lillard, 2017).

Over more than a century, this conception of space has remained strikingly consistent. From the first Casa dei Bambini to contemporary Montessori schools around the world, the prepared environment is recognizable by its calm beauty, open invitation to purposeful movement, scientifically designed materials, and confident trust in the child's unfolding development (NAMTA, 1990; NAMTA, 2001; Stephenson, 2013). It is a living testament to Montessori's belief that education is not transmission but a natural human process aided by the environment. When the world is prepared with intelligence, humility, and grace, the child's work becomes visible, and human potential reveals itself quietly, through hands engaged in meaningful action.

Montessori understood that this visible work of the hand and mind serves a deeper unfolding: the child's preparation for their **cosmic task**, the unique and meaningful contribution each human is called to make within the larger order of life. In the prepared environment, the child learns not only skills but orientation, discovering themselves as capable actors within a living universe guided by interdependence and purpose (Montessori, 1949/1995; Montessori, 1967). Through real work, freely chosen concentration, and responsibility for the community and natural world, the child begins to sense that they are part of a "great cosmic laboratory," entrusted with creative participation in the ongoing evolution of humanity (Montessori, 1967; Montessori, 1948/1995). As independence, moral awareness, and joyful effort take root, the child gradually recognizes that their developing abilities are not solely for personal advancement, but for service, stewardship, and contribution to the greater human story. The prepared environment therefore supports not only self-construction, but world-construction: a quiet initiation into the profound dignity of having a place and a purpose in the universe.

Chapter 9
The Child's Cosmic Task

In Montessori's vision, the education of the child extends beyond skill acquisition and intellectual development to a profound moral and cosmic orientation. She believed that every human being, from birth, carries within them the potential for a cosmic task: a unique contribution to the unfolding story of life and civilization. Montessori wrote that the child "comes into the world endowed with a mysterious potential which guides him to construct the universe" in both the material and spiritual sense (Montessori, 1949/1995). Human development, therefore, is not accidental or utilitarian; it is purposeful, directed toward participation in the harmony, progress, and common good of humanity.

Montessori's insight into the cosmic task emerged through observation and scientific study. Influenced by her medical work, anthropological research, and philosophical reflections on evolution and human culture, she recognized that all forms of life have roles in maintaining the balance and advancement of the world (Montessori, 1917/1992; Montessori, 1948/1995). Bees pollinate, earthworms aerate the soil, plants transform sunlight into nutrients; each element serves a necessary function. Children, too, she realized, are born with an inner directive energy that drives them toward meaningful contribution. "The child is truly a miraculous being," she observed, "and this should be felt deeply by the educator" as they support the child's gradual discovery of purpose (Standing, 1957).

This cosmic orientation becomes particularly clear in what

Montessori called Cosmic Education, introduced in the elementary years and grounded in a unifying view of the universe. Through grand narrative, integrated studies, and an emphasis on interdependence, children encounter the origins of life, the emergence of human culture, and the interwoven systems that sustain our world (Duffy & Duffy, 2002; Lillard, 2017). Montessori believed that offering the "vision of the whole universe" nourishes moral imagination and responsibility (Montessori, 1948/1995). By understanding how "all things are part of the universe and are connected," the child develops gratitude for past contributions and gains clarity about their place in the future (Montessori, 1967).

Importantly, Montessori placed cosmic work not solely in the realm of future adulthood, but within childhood itself. Children contribute in the present moment: they refine themselves, cultivate independence, build social harmony, and care for their environment. Practical life exercises, stewardship of plants and animals, collaborative classroom culture, and opportunities for leadership all serve as early expressions of civic and cosmic responsibility (Stephenson, 2013; McTamaney, 2009). In cultivating order, empathy, attention, and perseverance, the child develops what Montessori saw as the interior conditions necessary to fulfill their cosmic task with dignity and courage.

Within this framework, education becomes a preparation for purposeful participation in the world, not competition or compliance. Montessori's approach assumes trust in the child's inner guide and places the adult in the role of servant-leader, preparing environments that allow the child to discover not only knowledge, but meaning. As children learn to act with agency and responsibility, they begin to experience the profound truth at the heart of Montessori's method: each life matters, each person belongs, and each contribution ripples outward into the fabric of humanity. In recognizing their cosmic task, the child steps into a hopeful and humane vision of the future, carrying within them the seeds of peace, creativity, and service.

As we turn toward the intersection of Montessori education and emerging competencies for the future, including those shaped by artificial intelligence and digital systems, it is important to ground our inquiry in the universality and diversity of Montessori practice. The cosmic task reminds us that every child, community, and culture brings its own gifts and conditions, and that environments must adapt to context while holding fast to core developmental principles. Montessori herself emphasized responsiveness over rigidity, innovation guided by observation, and fidelity not to materials alone, but to the unfolding of human potential. With this spirit, we now consider how Montessori environments prepare children for participation in a rapidly evolving world.

Chapter 10

The Montessori Materials

As we explore the intersection of Montessori education and AI competencies, it is essential to acknowledge the diversity and adaptability within Montessori environments across the globe. While Montessori has a recognizable pedagogical structure and philosophy, its implementation reflects variation in history, culture, geography, resources, and community identity. Educators trained through organizations such as the Association Montessori Internationale (AMI) and the American Montessori Society (AMS) are introduced to a core sequence of scientifically designed materials that anchor developmental experiences in each plane. Yet access to these materials is not universal.

Around the world, Montessori guides prepare environments using a spectrum of approaches: from fully equipped classrooms in established schools to community-built environments in rural and underserved regions, from handcrafted materials to teacher-made adaptations shaped by observation and necessity. What unites these environments is not uniformity of objects, but commitment to the child's agency, concentration, dignity, and capacity to contribute to the world. In this light, the exploration of Montessori and AI competencies must honor both standardization where it supports integrity and innovation where it supports equity, relevance, and the child's cosmic task.

Montessori is practiced in diverse settings globally, and many communities lack access to traditional Montessori suppliers. In response, educators, caregivers, and communities have adapted by creating their own versions of key materials using locally sourced or repurposed resources, such as wood, metal, fabric, and sustainable plastics. These adaptations are not only practical but also reflect the ingenuity, cultural expression, and resourcefulness of those who implement Montessori principles in varied environments (Montessori, 1948/1995; Standing, 1957; NAMTA, 2001; Stephenson, 2013).

Before we present the specific materials used in this research, it is important to clarify that the materials referenced here are based on the American Montessori Society suggested materials list (AMS SML): *"AMS considers the use of Montessori materials essential for quality implementation. To support this, the AMS School Accreditation Commission and AMS Teacher Education Action Commission provide lists of recommended materials for each Montessori classroom level: https://amshq.org/educators/start-a-montessori-school*

American Montessori Society - Teacher Education

CORE MATERIALS AND ACTIVITIES FOR INFANT AND TODDLER PROGRAMS

The intent of this checklist is to provide Montessori teacher education programs with a list of developmentally-appropriate materials and activities that very young children ages birth to 3 years are generally expected to have available in their classrooms. This list is comprehensive but not exclusive or prescriptive and teachers may also present other activities.
Not all of these activities are expected to be out at any given time, but a wide variety of them should be provided at some point throughout the year. It is particularly important to consider cleanliness and safety with this age group, including awareness of potential choking hazards. As such, it is helpful for the adults to have a choke tube available to test for these hazards.

** Starred items indicate additional suggested materials or activities that may not be available at a school site, but will contribute to a new teacher's understanding of Montessori curriculum.*

	INFANTS – Birth to 18 Months
	Development of Movement: Control and Coordination
	Floor space – open areas for free movement
	Low climbing obstacles such as pillows and raised cushions
	Wall mirror – placed at floor level
	Floor beds
	Child-sized furniture, including tables and chairs of differing sizes, weaning chairs for younger infants, armless chairs for older infants.
	Bar for pulling up
	Free movement activities
	Rhymes and Singing with accompanying movements
	Maximum Effort activities
	Push and Pull toys
	Stair with rails
	Slide
	Inclined Plane
	Outdoor movement experiences
	Wheelbarrows and/or Wagon
	Sensori-Motor Development
	Black and White contrast mobile
	Variety of visual mobiles (different colors, different shapes)
	Variety of grasping (and kicking) mobiles (different colors, shapes, textures, and sounds)
	Ribbon and Bell
	Shape Sorting objects (Simple - no more than three shapes)
	Rattles
	Bell
	Scarves

Infant Toddler (0-3) Materials List - Page 1

	INFANTS – Birth to 18 Months
	Sensori-Motor Development, continued
	Soft Toys
	Nesting Objects
	Object Permanence activities
	Blocks
	Objects in a Basket for exploration
	Musical Instruments (Drum, Rhythm Sticks, Shaker Eggs, etc.)
	Music Box
	Sound System and Recordings
	Practical Life (Infants) **Care of Self, Care of the Environment, Grace and Courtesy**
	Many of the items listed in this section are to be introduced and modeled by the adult. These behaviors are not expected to be done consistently or independently by the child.
	Care of Self
	Bathing
	Changing a Diaper
	Floor mattresses or futons for sleeping
	Use of Handkerchief/Tissue
	Use of Toilet
	Washing Hands
	Washing Face
	Dressing and Undressing
	Feeding the Baby
	Self-Feeding
	Care of Environment
	Replacing materials on the shelf
	Drying spilled water
	Setting the table
	Serving oneself
	Watering plants
	Grace and Courtesy
	Greeting and saying goodbye
	Gentle touch with others
	Respecting work and space of others

2

Infant Toddler (0-3) Materials List - Page 2

	INFANTS -- Birth to 18 Months
	Language
	Communication experiences
	Vocabulary experiences
	Expressive and Receptive language experiences
	Visual Tracking experiences (Ball Tracker, Bead Maze, Mobiles, etc.)
	Puzzles with 1-2 pieces
	Books
	Songs
	Reciprocal vocalization
	Verbalization and Conversations (with the child)
	Naming Objects in the Environment
	Pictures of authentic activities for conversation
	Rhymes and Songs with actions
	Sign Language
	Vocabulary picture cards

	TODDLERS 18-36 Months
	Sensori-Motor Development
	Development of Coordinated Movement
	Space for gross motor movement and free movement activities
	Stair with Rails
	Climbing Objects (for climbing over and through)
	Walking in a Circle
	Balance Beam
	Bar from which to hang
	Rocking Horse/Boat/Chair
	Vestibular Board (e.g. sit and spin)
	Riding toys (e.g. tricycles)
	Maximum Effort activities (Carrying and pushing heavy objects)
	Outdoor Movement activities (walks, varying terrains, etc.)
	Outdoor Swing
	Slide
	Development of Motor Skills (Fine Motor Development
	Sequence of Transfer activities (large object transfer, small object transfer)
	Sequence of Scooping/Spooning activities
	Sequence of Squeezing activities (sponges, baster, tongs)
	Sequence of Stringing activities (Ring Posts, Beads and Posts, large bead stringing)
	Sequence of Twisting activities
	Sequence of Pouring activities
	Sequence of Rolling and Folding activities

IT Activities and Materials 2020 3

Infant Toddler (0-3) Materials List - Page 3

	TODDLERS 18-36 Months
	Sensori-Motor Development
	Sequence of Opening and Closing activities (opening and closing drawers, opening and closing containers)

	Practical Life (Toddlers)
	Personal Care and Health
	Dressing and Undressing
	Using the Toilet
	Washing Hands
	Simple Dressing Frames (zipper, button, snap, buckle, velcro)
	Bathing the Doll
	Object washing (tables, chairs, simple objects)
	Folding and Putting Clothes Away
	Taking off and Putting on a Coat
	Taking off and Putting on Shoes
	Using a Tissue/Wiping Nose
	Brushing Teeth *
	Combing Hair *
	Care of Environment
	Using a rug or mat
	Using child-sized furniture
	Replacing materials on the shelf
	Setting the Table
	Wiping the Table
	Sweeping and Dusting
	Cleaning up spills
	Mopping the floor
	Washing Dishes
	Scrubbing objects (tables, cloths, baby doll)
	Washing Windows
	Caring for Pets (providing food and water)
	Caring for Plants (watering, polishing leaves)
	Flower arranging
	Folding
	Shoveling and Raking
	Planting seeds & plants
	Picking up sand, mulch, rocks
	Outdoor activities to care for the environment

4

Infant Toddler (0-3) Materials List - Page 4

	Practical Life (Toddlers)
	Food Preparation
	Washing fruits and vegetables
	Cutting fruits, vegetables, cheeses, eggs, etc.
	Mixing Ingredients
	Spreading (nut butters, cream cheese,etc)
	Preparing drinks (making juice, tea, etc.)
	Serving Food (with hands, scoops, spoons, tongs, etc.)
	Clearing the Table
	Washing Dishes
	Development of Grace and Courtesy
	Greeting and saying goodbye
	Asking for help
	Giving assistance
	Respectful requests and polite conversation (Use of please, thank you, excuse me)
	Gentle touch
	Respecting work and workspace
	Developmental Art Materials and Activities
	Clay and/or Play Dough
	Chalkboard
	Vertical Easel
	Painting
	Crayons
	Simple Scissors
	Glue with small brush or glue sticks

	Sensori-Motor Development (Toddlers)
	Stacking/Nesting Boxes/Nesting Objects
	Object Permanence Boxes
	Geometric Shape Box / Imbucure Box with Geometric Shapes / Shape Sorter
	Puzzles (including a geometric shape puzzle)
	Sorting
	Color Matching
	Music
	Sounds in the Environment
	Voice: Tone and Volume
	Simple Sound Matching (4 Pairs)
	Percussive Instruments (e.g. maracas, small floor drums, bells, rain sticks)
	Music Box
	Mystery Bag
	Texture and Temperature in the Environment
	Sandbox

IT Activities and Materials 2020 5

Infant Toddler (0-3) Materials List - Page 5

	Sensori-Motor Development (Toddlers)
	Rough and Smooth Textures
	Smelling and Tasting experiences (e.g. food, flowers, grinding spices)
	Outdoor experiences in nature

	Language (Toddlers)
	Development of Oral Language
	Opportunities for receptive and expressive language (conversations and exploration)
	Oral language games
	Singing and storytelling
	Stories with finger plays, felt boards, relevant objects
	Books (Objects, Actions, Stories)
	Rhyming Games
	Reciprocal vocalization
	**Non-working telephone*
	Development of Vocabulary and Classification
	Naming objects in the Environment
	Naming objects (baskets of objects, picture cards)
	Object to Object matching
	Object to Picture matching
	Picture to Picture matching
	Vocabulary Cards
	Farm
	Sequence of Knobbed Puzzles (Large knobbed puzzles, Small knobbed puzzles)
	Categorization of Objects and Pictures (Fruits and Vegetables, Cats and Dogs, etc)
	Matching: Pictures of Objects that Go Together (i.e. spider and spider web)
	Mystery Bag
	Objects for "I Spy" with initial sounds

The Adults in the Environment
It is important to recognize that the qualities, skills and behaviors of the adults responsible for meeting the needs of these young children is incredibly important during this most important stage of self-construction. This is the crucial time in which children are absorbing experiences, making early decisions about the kind of persons they are, and about the world. They are developing basic trust in themselves and in their environment, which will affect them for the rest of their lives.

Infant Toddler (0-3) Materials List - Page 6

American Montessori Society - Teacher Education

CORE MATERIALS AND ACTIVITIES FOR EARLY CHILDHOOD PROGRAMS

The intent of this checklist is to provide Montessori teacher education programs with a list of traditional developmentally appropriate activities that Montessori teachers of children ages 2 1/2 through 6 years are generally expected to be able to present in their classrooms. This list is comprehensive but not exclusive or prescriptive and teacher education program instructors may also present other activities.

** Starred items indicate additional suggested materials or activities that may not be available at a school site, but will contribute to a new teacher's understanding of Montessori curriculum.*

	Practical Life
	Refinement of Movement
	Walking on a Line (incl. carrying objects while walking)
	Silence Games
	Sequence of Scooping/Spooning activities
	Sequence of Squeezing activities (e.g. sponge/baster/clothespins/tongs/tweezers)
	Sequence of Threading activities
	Sequence of Twisting activities (bottles & caps, nuts & bolts, etc.)
	Sequence of Pouring activities (dry / water / various size, shape containers / with tools)
	Sequence of Folding activities
	Care of Environment
	Use of rugs/mats (e.g. rolling, carrying, avoid walking on work rugs)
	Use of child size furniture (e.g. moving tables and chairs)
	Sweeping and dusting (e.g. broom, dustpan & brush, table brush & pan, duster)
	Mopping the floor (e.g. mop and bucket)
	Scrubbing surfaces (e.g. table / chair / cloths)
	Washing windows, using squeegee
	Washing dishes
	Polishing e.g. (mirror/wood/metal)
	Sequence of sewing activities (e.g. sewing on a button)
	Arranging flowers
	Caring for plants (watering / polishing leaves)
	**Caring for small animals (providing food and water)*
	** Sequence of carpentry activities (Use of vises/ clamps / hand saw/ pliers / level/ goggles / gloves / screwdriver & screws/ hand drill / hammer / nails / wood / sandpaper)*
	** Caring for the outdoor environment (e.g. brooms, rakes, garden tools)*
	**Gardening (Sowing seeds, weeding, harvesting)*
	Grace and Courtesy
	Greetings and introductions
	Respectful and polite conversations (e.g. interruptions, apologies, requests)
	Manners for meals and group settings
	Classroom procedures

EC Activities and Materials 2020 1

Early Childhood (3-6) Materials List - Page 1

	Practical Life
	Personal Care and Health
	Storing personal clothing and belongings (e.g. cubbies, hooks, coat hangers)
	Dressing Frames – Buttons, zipper, buckles, lacing, bow-tying
	Dressing and Undressing (e.g. putting on jacket and shoes)
	Cleaning and polishing shoes
	Nose-blowing (tissue / mirror)
	Toileting and use of the bathroom facilities
	Hand-washing (e.g. table set up w/large bowl & pitcher, etc., and use of sink with faucet)
	* *Individual personal care items (e.g. combs, toothbrushes)*
	Food Preparation
	Washing fruits and vegetables
	Cutting fruits, vegetables, cheese, eggs, etc.
	Preparing drinks (juice, tea, etc.)
	Serving food and drinks
	**Cooking and baking activities*

	Sensorial
	Cylinder Blocks (Knobbed Cylinders) - 4 blocks
	Tower of Cubes (Pink Tower)
	Broad Stair (Brown Stair)
	Long Rods (Red Rods)
	Knobless Cylinders - 4 boxes
	Color Tablets - Box I
	Color Tablets - Box II
	Color Tablets - Box III
	Geometric Cabinet w/ Demonstration Tray and 3 sets of cards
	Geometric Solids and Bases
	Constructive Triangles (5 boxes): Triangle Box, Large Hexagon Box, Small Hexagon Box, Rectangle Box, Blue Triangles Box
	Binomial Cube
	Trinomial Cube
	Sound Cylinders (Sound Boxes)
	Montessori Bells Set
	Rough and Smooth Boards - 2 or 3 boards
	Touch Tablets (Rough & Smooth Tablets for matching and grading)
	Fabric Matching
	Mystery Bag (Stereognostic)
	Baric Tablets (Weight)
	Thermic Tablets

2 EC Activities and Materials 2020

Early Childhood (3-6) Materials List - Page 2

	Sensorial
	Thermic Bottles
	Smelling Jars
	**Tasting Activities*
	** Sensorial Vocabulary Labels (names of colors, shapes, sizes, etc.)*

	Math
	Number Rods (Red and Blue Rods) and Numerals
	Sandpaper Numerals
	Spindle Boxes
	Cards and Counters (Numerals and Counters)
	Memory Game
	Intro to Decimal Quantity (One Tray) and Symbol
	Function of the Decimal System (Nine Tray)
	Golden Bead Material for Addition, Multiplication, Subtraction, Division (Sufficient for 45 layout)
	Large Decimal Numeral Cards (1 set)
	Small Decimal Numeral cards (3 sets)
	Short Bead Stair
	Teen Boards and Beads (Séguin Board I)
	Tens Boards and Beads (Séguin Board II)
	Hundred Board
	Short Bead Chains w/ squares and numerals (Square Chains)
	Long Bead Chains w/ cubes and numerals (Cube Chains)
	Addition Snake Game
	Addition Strip Board and Tables
	Multiplication Bead Bars
	Multiplication Board and Tables
	Subtraction Snake Game
	Subtraction Strip Board and Tables
	Addition Finger Charts and Box of Equations
	Multiplication Finger Charts and Box of Equations
	Subtraction Finger Charts and Box of Equations
	Unit Division Board and Box of Equations
	Division Finger Charts
	Fraction Circles (Metal Fraction Insets)

	Language
	Child library w/ fiction and non-fiction picture books
	Oral language games, storytelling, listening, self-expression, conversations, etc.
	Sensorial oral vocabulary language games
	Phonemic awareness games (e.g. I-Spy)
	Oral vocabulary enrichment activities (reflecting cultures of the children)
	Picture/Object Matching and/or Picture/Picture Matching

EC Activities and Materials 2020 3

Early Childhood (3-6) Materials List - Page 3

	Language
	Sorting objects and/or pictures into categories
	Sandpaper Letters (Lower case – cursive or print)
	Double Sandpaper Letters (English phonograms)
	Sand Tray
	**Upper case Sandpaper Letters (cursive or print)*
	Large Movable Alphabets (several)
	Metal Insets w/ pencils, pencil holders, trays, paper
	Small chalkboards for handwriting & supplies *(*white boards)*
	** Large Name Card for each child in the class*
	**Activities for stroke and placement of letters*
	**Activities for handwriting letters and words*
	Object Box I: Introduction to reading short vowel phonetic words
	Object Box II: Introduction to reading words w/ phonograms (for English)
	Environment Label cards
	Action Command Cards
	Decoding short vowel words (e.g. flash cards, picture/word booklet, picture cards with labels)
	**Simple phonetic reading such as a phonics-based series of reading books*
	Puzzle Words (Sight Words)
	Phonogram reading cards and booklets (for English)
	Silent 'e' cards (for English)
	** Small printed alphabets in two colors for use with phonogram materials*
	Three-Part Cards – (organized by category) for blending and fluency
	Mini Environment (e.g. Farm) for introduction to parts of speech
	Grammar Symbols
	Function of Words - introductory lessons
	Variety of reading material for emerging fluent / beginning readers (fiction and non-fiction)
	**Activities to encourage written self-expression and simple research*

	Cultural (Geography, History, Natural and Physical Sciences)
	Geography
	Land/Air/Water – 3 containers
	Land and Water Form Trays w/ Pitcher
	Three-Part Cards and Definitions for Land and Water Forms
	Globe of Land and Water (Sandpaper Globe)
	Globe of the Continents (Colored Globe)
	**Flattened Globe Demonstration*
	Puzzle Map - World Parts (Continents) + Control map, and Circle for tracing
	Puzzle Map – North America + Control map
	Puzzle Map – South America + Control map
	Puzzle Map – Europe + Control map
	Puzzle Map – Asia + Control map

4 EC Activities and Materials 2020

Early Childhood (3-6) Materials List - Page 4

	Cultural (Geography, History, Natural and Physical Sciences)
	Puzzle Map – Africa + Control map
	Puzzle Map – Australasia (Oceania) + Control map
	Puzzle Map – Children's home country (e.g. USA, Mexico, Canada, Japan, China) + Control map
	Tools and supplies for reproducing maps and flags
	Continent Folders / baskets (objects / pictures representing the continent/country)
	**Animals of the continents*
	Flags of a variety of countries
	**Map of Oceans*
	**Introduction to Solar System (Sun and Planets)*
	Science
	Living/Non-Living Classification
	Vertebrate/Invertebrate Classification
	Animal/Plant Classification
	**Classified Picture Cards – Animals (different characteristics e.g. hooves, horns, zoo, domestic)*
	**Classified Picture Cards – Plants (different characteristics)*
	**Puzzles for parts of mammal, bird, reptile, amphibian, fish*
	Parts of a Mammal (three-part cards & definitions)
	Parts of a Bird (three-part cards & definitions)
	Parts of a Reptile (three-part cards & definitions)
	Parts of an Amphibian (three-part cards & definitions)
	Parts of a Fish (three-part cards & definition materials)
	**Parts of an Insect (three-part cards & definitions)*
	Botany Cabinet and Cards (Leaf shapes)
	Parts of a Tree (three-part cards & definitions)
	Parts of a Flower (three-part cards & definitions)
	Parts of a Leaf (three-part cards & definitions)
	Parts of Root (three-part cards & definitions)
	**Life cycles – plants / animals*
	**Plant experiments*
	**Rocks / minerals / shells*
	Physical Science
	**Sink or Float activity*
	**Magnetic or Non-magnetic activity*
	**Experiments for physical science discovery*
	** Simple machines*

Early Childhood (3-6) Materials List - Page 5

	History
	Calendar Activities (e.g. day / date / month etc.)
	Months of the Year cards (used for child's birthday celebrations)
	Materials for seasons (e.g. climate / weather / clothing / food / activities / change in foliage)
	** Timer Devices (hourglass / variety of clocks / variety of calendars / sundial)*
	** Timeline of child's day*
	** Timeline of child's birthdays*
	** Timeline of child to adult*
	**Artifacts from different eras (e.g. toys / transportation / technology / clothing)*

	Creative Arts
	Arts and Crafts
	Clay & cutting tools
	Variety of Tools for Drawing (e.g. pencils, crayons, brushes, pastels, chalk)
	Varied types of paper (colors, sizes, textures
	Scissors & materials for cutting (also tearing)
	Glue for collage
	Painting supplies (easel / tempera paint / water colors)
	Reproductions of famous artworks to introduce artists and art from varied cultures
	Music & Dance
	Source for recorded music
	Rhythm Instruments (e.g. triangle, tambourine, maracas, cymbals, shakers, sticks, drums, etc.)
	**Models of musical instruments for matching and naming*
	**Pictures of composers and sample of compositions from varied cultures*
	**Materials for dancing (e.g. ribbons and scarves)*

Early Childhood (3-6) Materials List - Page 6

American Montessori Society – Teacher Education

CORE MATERIALS AND ACTIVITIES FOR ELEMENTARY PROGRAMS

The intent of this checklist is to provide Montessori teacher education programs with a list of traditional, developmentally appropriate materials and activities that Montessori teachers of children ages 6 through 12 years are generally expected to have available in their classrooms. This list is comprehensive, but not exclusive or prescriptive and teacher education program instructors may also present other activities.

If a material is specifically for the EI or EII level, a mark has been placed in the appropriate column. "Transition" indicates that the material or activity is often used in classrooms at both age levels.

**Starred items indicate additional suggested materials or activities that may not be available at a school site, but will contribute to a new teacher's understanding of Montessori curriculum.*

	Cultural Knowledge (Geography, History, Science)	E I	Transition	E II
	GEOGRAPHY - Political			
	Puzzle Maps of Continents/Control Maps	x		
	Flags of The World and Stands	x		
	Nomenclature of the Flag	x		
	Physical and Political Pin Maps	x	x	
	GEOGRAPHY- Physical			
	Land Forms	x		
	Classified Nomenclature for Geography	x		x
	Experiments / Demonstrations to prepare for the Impressionistic Charts	x		x
	Geography Impressionistic Charts:			
	- The Universe, The Solar System, The Earth	x		x
	- Solar Energy and The Earth	x		x
	- Movements of the Earth/Consequences	x		x
	- The Atmosphere and Its Phenomena		x	x
	- The Work of the Wind		x	x
	- The Hydrosphere and Its Phenomena		x	x
	Format for In-depth Study of a Country*		x	x
	Economic Geography Charts		x	x
	Interdependence Activities*		x	x
	Production/Consumption -Maps and graphs		x	x

EL Activities and Materials 2020

Elementary (6-12) Materials List - Page 1

	HISTORY			
	Concept of Time	x		
	Days, Weeks, Months, Year and Seasons	x		
	Personal Time Lines	x		
	Telling Time - Clock	x		
	BCE/CE and/or BC-AD Time Line w/ Golden Bead materials, Century labels, and representative pictures	x		x
	Long Black Line (timeline) - Coming of Humans	x	x	
	Clock of Eras	x	x	
	Timeline of Life with Control and Mute charts	x		x
	Fundamental Needs of People - chart	x	x	
	Timeline of Peoples w/ cards from time periods and civilizations		x	x
	Migration charts			x
	Timelines for the Study of Civilizations		x	x
	Timeline of child's own country			x
	LIFE SCIENCE			
	Timeline of Life	x		x
	Zoology			
	Vertebrate-Invertebrate Classification - pictures for sorting and further classification	x		
	External Parts of Vertebrates - Wall Charts, 3 Part Cards, Booklets	x		
	Internal Parts of Vertebrates - Wall charts, 3 Part Cards, Booklets	x		
	Animals First Knowledge - Pictures, labels, story cards	x		
	Animals - First Knowledge - Questions & Answers	x		
	Animal Kingdom - Classification Chart	x		
	Invertebrates - Classification	x		
	Animal Kingdoms - Chart	x		x
	Animal Kingdoms - Domains of Life		x	x
	Comparative Functions of Animal Systems			x
	Great River (Study of human body)			x
	Tree of Life - Chart			x
	Botany			
	Botany Nomenclature: Plant, Roots, Stem/Trunk, Leaves, Flowers, Fruit, Seed	x		
	Botany Impressionistic Charts	x		
	Plants First Knowledge Stories: Pictures, labels, story cards	x		
	Plant Classification Charts		x	x
	Plants First Knowledge: Questions & Answers		x	x
	Advanced Botany Nomenclature			x
	Materials for Botany Experiments	x		x

EL Activities and Materials 2020

Elementary (6-12) Materials List - Page 2

Geometry		E I	Transition	E II
	Geometric Cabinet	x		
	Constructive Triangles	x		x
	Geometric Solids	x		
	Geometry Stick Material and Board	x		x
	Classified Nomenclature	x		x
	Equivalency Material I - Square divided into rectangles and triangles	x	x	
	Insets of Equivalency - for discovering area formulas			x
	Yellow Area Material			x
	Stand for Height of Triangle			x
	Theorem of Pythagoras plates			x
	Circle – Circumference and Area Materials		x	x
	Volume – Solid and Hollow Blue Shapes			x
	Volume box with 250 cubes			x
	Volume box with 1000 cubes			x
	Five Prisms			x
	Montessori Protractor (Instrument for Measurement of Angles)	x	x	x

Language		E I	Transition	E II
	History of Language*	x		x
	Comprehensive Phonetic Reading Program	x		
	Printed Bi-color Movable Alphabets	x		
	Phonetic Word Building Activities	x		
	Sandpaper Letters for Handwriting	x		
	Words to Pictures Cards	x		
	Sentences to Pictures Cards	x		
	Puzzle Words (Sight Words/Non-phonetic)	x		
	Command Cards: 1, 2, 3 Actions	x		
	Word Studies Activities	x		x
	Mini-Environment and labels e.g. Farm	x		
	Grammar Boxes	x		x
	Grammar Symbols and Activities	x		x
	Sentence Analysis Charts and Boxes with Arrows for Sentence Analysis/Reading Analysis/Sentence Construction/ Logical Analysis	x		x
	Access to a Library for research and pleasure reading (classroom or school)	x		x
	Material for Irregular Verbs	x		x
	Verb Forms		x	x
	Conjugation of Verbs		x	x

EL Activities and Materials 2020

Elementary (6-12) Materials List - Page 3

Practical Life Practical Life in an elementary environment is a preparation for life, when the classroom expands outwards to the community and the world. The elementary environment should make available materials, supplies and activities that support a cosmic education.	**E I**	**Transition**	**E II**
Refinement and Control of Movement Movement and physical education activities, such as intentional control of movement, mindfulness, meditation, and refined movement with materials	x		x
Social Relations Awareness and care of school and global community Grace and courtesy Education for peace Social justice Diversity Inclusion Community agreements Community meetings Service projects	x		x
Care of the Shared Environment Care of the classroom environment - dusting, cleaning, polishing, mending Care of plants and animals Tools for outdoor work "Going Out" activities, field experiences/studies	x		x
Personal Care and Care of Others Supplies for personal hygiene and self-care. Food preparation and Cooking	x		x
Creative Arts Ample supplies to encourage spontaneous creative expression and exploration			
Music Drama Crafts e.g. sewing, art, carpentry,	x		x

EL Activities and Materials 2020

Elementary (6-12) Materials List - Page 4

American Montessori Society – Teacher Education

CORE MATERIALS AND ACTIVITIES FOR SECONDARY PROGRAMS

These materials and activities are for Secondary I & II grade levels (7th – 12th grade). Their placement and how they are used in the classroom/school will depend on whether there is one large open space for the program or individual classrooms.

The intent of this list is to provide AMS Montessori teacher education programs with a list of developmentally appropriate materials and activities that credentialed Montessori teachers of adolescents ages 12 through 18 years (grades 7-12) are advised to have available in their classrooms. This list is not exclusive or prescriptive and teacher education program instructors may present other activities.

* *Starred items indicate suggestions that may not be available, but will contribute to the Montessori program.*

	Physical Environment
	Space and furnishings to accommodate whole group and small group work
	Variety of comfortable personal work spaces
	Designated storage space for each student's personal and work items (e.g. lockers)
	Restrooms nearby - stocked with personal hygiene products
	Mirror (*Optimal: full length mirror* - an adolescent's entire body is undergoing significant physical changes)
	Easy access to filtered drinking water
	* *Nearby kitchen for use by students*
	* *Science sink and lab area*
	* *Large windows and natural lighting*
	**Access to nearby outdoor work space*
	**Opportunities for outdoor/land experiences*

SEC Materials and Activities 2021

Secondary (12-18) Materials List - Page 1

	Basic Equipment and Supplies
	Handbook for Secondary students & parents describing specific secondary class policies, procedures and curriculum (either separate from or included as part of whole school handbook)
	Classroom computers and/or student laptops/tablets or similar
	Printer
	Dry erase markers & board(s) - large & individual size as appropriate
	Office equipment - accessible to students e.g. scissors, hole punch, pencil sharpener, stapler
	Office supplies and art materials - accessible to students e.g. pens, regular & colored pencils, markers, paints, poster board, glue
	Paper - ruled, unruled, graph, color
	Cleaning supplies for care of the environment by the students e.g. broom, dustpan, rags, soap, cleaners (non-toxic)
	First Aid supplies: Cold pack, band-aids, ointments, etc.
	* *Copier*
	* *Tools, equipment, materials for real-life experiences/activities per each program*

	Curriculum and Resources
	The *specific* subject area resources are as appropriate to a MS/HS student body and state/country requirements
	Language: writing for many purposes, grammar, culturally relevant literature, varied genres
	Math: appropriate manipulatives especially for initial presentations
	Sciences: materials for experiments and hands-on work – microscopes, properly stored chemicals, etc
	Social Studies: primary resources; varied secondary texts that present multiple perspectives
	Additional or world languages
	Intra/Interpersonal: sources for positive identity development, communication skills, conflict resolution, goal setting, dealing with stress, sex education, etc
	Technology: thoughtful and ethical use of technology
	The Arts: variety of offerings, separate from or embedded in courses e.g. performance, visual, musical, pottery, dance, sculpting, graphic design, etc
	Multiple information resources e.g. wi-fi access, books, charts, graphs
	Variety of dialogue resources e.g. texts, visuals, videos, music selections
	Variety of media supporting all areas of the curriculum

SEC Materials and Activities 2021

Secondary (12-18) Materials List - Page 2

	Secondary I or Secondary I-II AMS credentialed teacher
	Student body consists of multi-age grouping of students in the Third Plane of Development (two grade levels – or more – within 7th - 12th grade level range)
	Core subject multi-age grouping of students in the Third Plane of Development (minimum of two grade levels in at least two core subjects if core subjects are separated)
	Easy access to traditional Montessori materials and charts for transition lessons
	Resources for short term electives for adolescents to explore interests such as pottery, yearbook, cooking, creative writing, videomaking, etc.
	Real-Life Work Experience: * *Operating a small business* * *Gardening or farm work* * *Care of animals* * *Food preparation and cooking* * *Service learning* * *Business internships* * *Infant simulator doll*
	Outdoor/indoor physical activities and games
	Other subjects and electives as relevant, or as needed to meet local/state/country requirements

Key Activities	
	Daily routines that are consistent, simple, and dependable e.g. most if not all days follow the same schedule
	Systems in place e.g. how & when to problem solve whole class issues; turn in work; what must occur before a retest, etc.
	Orientation at beginning of school year (inclusive with at least two grade levels) e.g. establish classroom systems; team building activities; integration of new students
	Regular whole group student-led community meetings (inclusive with at least two grade levels)
	Regular opportunities for small group work (inclusive with at least two grade levels)
	Regular opportunities for choice in academic work
	Regular opportunities for students to practice time management
	Regular opportunities for student self-assessment

SEC Materials and Activities 2021

Secondary (12-18) Materials List - Page 3

Key Activities	
	Regular opportunities for structured student discourse/dialogue (inclusive for at least two grade levels. There are several dialogue guidelines from which to choose for structured dialogue)
	90- to 120-minute periods of uninterrupted work time
	Regular protected time for solo/personal reflection/mindfulness
	Regular opportunity for physical exercise
	Daily student care of their environment
	Opportunities for partnering with families e.g. student-lead parent conferences
	* *Field studies*
	**Opportunities for micro community living 24/7*
	**Opportunities for time on the land to connect with the earth*

SEC Materials and Activities 2021

Secondary (12-18) Materials List - Page 4

While they serve as a structured reference, it is not an exhaustive or exclusive compilation of materials used in Montessori settings worldwide. This research does not include lists from AMI, independent Montessori programs, homeschooling communities, or alternative Montessori material manufacturer. It is crucial to understand that the materials listed should be viewed as conceptual placeholders rather than prescriptive tools.

If you or your community have created adaptations of these materials, whether by substituting different materials, altering dimensions, or making them more culturally relevant, your innovations are fully acknowledged and valued within this framework. Montessori education is inherently adaptive, and the use of alternative materials aligns with its core principle of following the child and meeting learners where they are.

For example, if a traditional Montessori classroom includes a Pink Tower, but an educator or parent has designed an equivalent using a different material or format while keeping the structural elements, the function and purpose of the material remain intact. The essence of Montessori education is not tied to rigid physical materials but to the intentionality behind their purpose and their use, supporting children's cognitive, sensory, and motor development.

It is also important to recognize that Dr. Maria Montessori did not trademark her name or method. Instead, she made her method "Open Source," available to researchers, educators, and communities worldwide. She encouraged continuous observation, adaptation, and innovation in response to children's evolving needs. The creation of materials was central to her approach, she designed learning tools based on direct observations of children, refining them based on their responses. This tradition of innovation remains essential in Montessori education today.

The purpose of this discussion is to reaffirm that embracing Montessori pedagogy does not mean strict adherence to a singular set of materials. Each community, school, and home-based learning environment is unique, and the materials used should reflect the needs, cultures, and available resources of those communities.

As you engage with conversations regarding the materials in this research, please view them as guiding principles rather than absolute requirements. Consider them starting points for inquiry and adaptation, rather than fixed mandates. If you have created alternative versions of any Montessori materials, whether for infants, toddlers, primary, elementary, or adolescent learners, identify their nearest functional equivalent within this framework and continue the work of fostering meaningful, developmentally appropriate learning experiences for children.

At the time of publishing, the American Montessori Society (AMS) has partnered with Fisher-Price to develop a Montessori-approved toy collection. This collaboration marks a significant step in expanding access to Montessori-aligned materials beyond the classroom, bringing key principles of child-led, hands-on learning into homes through widely available retail channels.

By embracing a broad and inclusive perspective, we honor the spirit of Dr. Maria Montessori: one that is innovative, and deeply responsive to the needs of every child and the times in which they live. Thank you for your commitment to creating rich, engaging, and accessible learning environments for all children.

It is in this spirit that the following mapping exercise was undertaken, to examine how Montessori materials, long grounded in developmental science and observation, align with the emerging competencies required in the age of artificial intelligence and other transformative technologies. Rather than reinvent the method, this work seeks to illuminate what is already present, revealing how Montessori education continues to prepare children not just for the world as it is, but for the world as it is becoming.

And at the very heart of this preparation stand the Great Lessons, the grand narratives that open the Montessori elementary curriculum and weave together the whole of human knowledge. They are not simply beautiful stories; they are the architecture of an interconnected curriculum, where every strand of mathematics, language, science, history, and the arts flows outward from these first great stories. In understanding the Great Lessons, we understand how Montessori education holds the whole universe in its hands, and invites the child to step into it.

Acknowledgment – I would like to extend my deep gratitude to the American Montessori Society for the Suggested Materials List made freely available on their website under the "Starting a Montessori School" section. This resource has been instrumental throughout my personal journey, as a Montessori teacher-in-training, as a guide, and later as the director of a school. It provided both clarity and confidence at every stage, helping to ensure that the prepared environment truly served the developmental needs of the child. I continue to value and reference this list as a foundation for thoughtful, authentic Montessori practice.

Chapter 11

The Great Lessons: The Whole Story for the Child

In the Montessori elementary classroom, the year begins not with a textbook or checklist but with a story, or rather, with a constellation of stories known as the Great Lessons. These sweeping narratives are told to ignite the imagination and awaken a child's sense of belonging.

They do not exist to deliver facts, but to stir wonder: Why does the universe exist? How did life arise? What makes humans unique? How did we begin to speak, to write, to measure, to calculate?

Maria Montessori captured this intention in *To Educate the Human Potential* (1948):
"Since it has been seen to be necessary to give so much to the child, let us give him a vision of the whole universe. The universe is an imposing reality, and an answer to all questions."

The Great Lessons provide that vision. They open a framework so vast that every subject, history, mathematics, science, language, art, finds its place. They invite children to build a map of knowledge in which all things are interconnected.

What follows is not material to be mastered in detail, but context. To understand how artificial intelligence and digital competencies can be meaningfully placed within the Montessori curriculum, it is essential first to glimpse the foundation on which everything rests.

The Great Lessons are that foundation. They orient the child to the universe, to life, to humanity, to language and numbers, and to the unfolding story of discovery and invention. They are not isolated "subjects" but invitations into a coherent vision of human development, one in which new strands, including AI and digital citizenship, can be integrated without distortion.

You need not study the Great Lessons chapter by chapter if your primary focus is AI. Simply know that this is the soil from which everything else grows.

The First Great Lesson: The Coming of the Universe and the Earth

The story begins before time as we know it. Silence, darkness, and then, a great unfolding. The universe expands, stars ignite, galaxies spin into being, Earth cools and solidifies, oceans form, and volcanoes breathe fire.

From this first tale flows a cascade of studies:

- **Astronomy**: stars, planets, galaxies, cosmic forces.
- **Physics**: matter, energy, laws of motion, gravity.
- **Chemistry**: atoms, molecules, elements, reactions.
- **Earth science**: geology, landforms, weather, plate tectonics.
- **Geography**: continents, oceans, ecosystems, mapping.
- **Time**: clocks, calendars, cosmic timelines.

The child hears not only the grandeur of cosmic creation but also the reminder: *You are made of this. The elements of your body were born in stars.*

The Second Great Lesson: The Coming of Life

When Earth settled, life appeared in the oceans. Invisible organisms transformed air and water, preparing the way for the great pageant of living forms: plants, fish, reptiles, birds, and mammals.

From this story unfold studies such as:

- **Biology**: classification of plants and animals, botany, zoology.
- **Evolution**: fossils, adaptation, natural selection.
- **Ecology**: food chains, interdependence, conservation.
- **Paleontology**: prehistoric life and ancient seas.
- **Environmental science**: human impact on ecosystems.

Children trace the long patience of life, learning that every creature, no matter how small, played a role in preparing the world for what would come next.

The Third Great Lesson: The Coming of Human Beings

When humans enter the story, the emphasis is not on conquest but on the gifts Montessori called the “great powers”: hands for work, minds for imagination, and hearts for love and community.

This lesson opens into:

- **Anthropology**: early humans, migrations, tools, shelters.
- **History**: ancient civilizations, cultural development.
- **Geography**: settlement patterns, maps of human movement.
- **Fundamental Needs of Humans**: food, clothing, shelter, defense, transportation, art, spirituality, love.
- **Sociology**: communities, governance, family structures.
- **Economics**: barter, trade, money, agriculture.

Here, the child sees not only what humans have made, but also the responsibilities that come with being human.

The Fourth Great Lesson: The Story of Language

Once survival was secure, humans shaped sound into meaning. Words emerged, followed by written signs, alphabets, and literature. Language became the thread that bound generations together.

From this lesson come explorations of:

- **Spoken language**: phonetics, etymology, linguistics.
- **Written language**: scripts, hieroglyphs, alphabets, calligraphy.
- **Grammar**: parts of speech, sentence analysis.
- **Literature**: myths, poetry, oral traditions, world literature.
- **Communication**: rhetoric, symbols, signs, storytelling.
- **History of writing**: from clay tablets to the digital word.

Language is revealed as both practical and poetic, a gift that carries memory across time.

The Fifth Great Lesson: The Story of Numbers

Alongside language, humans invented mathematics. First came marks on stone and tally sticks, then number systems and calendars, later algebra and geometry. Numbers helped us navigate seas, build temples, compose music, and eventually explore the universe itself.

Subjects that open from this lesson include:

- **Arithmetic**: place value, operations, fractions, decimals.
- **Geometry**: shapes, constructions, angles, measurement.
- **Algebra**: patterns, unknowns, relationships.
- **History of mathematics**: zero, number systems, mathematicians.
- **Applied mathematics**: architecture, engineering, art, music, navigation.
- **Statistics and probability**: data, logic, coding, information science.

Children discover that mathematics is not drudgery, but a human story, one of curiosity, problem-solving, and beauty.

The Sixth Story (Optional): The Story of Writing

In some traditions, the history of writing is given as a separate Great Lesson. It can lead to:

- **Calligraphy and handwriting**.
- **The history of books**: printing press, libraries, archives.
- **Storytelling**: as cultural memory and artistic expression.

The Great Lessons as a Cosmic Framework

Each Great Lesson is a key that unlocks entire fields of knowledge. They are told at the beginning of the year not as a "unit of study" but as a foundation for wonder. Every subject a child encounters afterward can be traced back to one of these stories.

Montessori called this **cosmic education** the weaving of all disciplines into a meaningful whole. Instead of fragmenting knowledge into isolated boxes, the child learns that all things are connected: the atom to the galaxy, the plant to the ecosystem, the human heart to the history of culture.

As the lessons unfold, the child comes to see that they themselves are part of this great story.

Closing Reflection

The Great Lessons are gifts, not just lessons. They are offered to the child with reverence, like torches lit at the beginning of a journey. Each child is invited to walk forward with curiosity and responsibility, knowing that the universe is not a backdrop but a living story in which they have a role.

Or, as Montessori reminded us in *To Educate the Human Potential* (1948): *"The universe is an imposing reality, and an answer to all questions."*

And so the Great Lessons open a horizon of wonder, a framework in which every discipline becomes alive and meaningful. They place the child within the cosmic story, not as a passive observer but as an active participant, capable of shaping the future. Yet this vision is only the beginning. If the universe is the stage, the child is the unfolding drama. It is from this foundation that we now step forward, turning from the great story of the cosmos to the equally profound story of human development, the ordered journey of growth through which each child comes to know themselves and their place in a rapidly changing world.

Chapter 12

The First Great Lesson and Its Pathways

The First Great Lesson is not a single story that ends with silence. It is a seed, and from it spring whole branches of knowledge. Children do not leave the tale of the universe behind; rather, they walk its pathways, astronomy, physics, chemistry, geology, geography, and the measurement of time, each one a door into deeper discovery.

Montessori Elementary Cosmic Curriculum (Post–First Great Lesson)

Astronomy

Astronomy connects the child with the cosmos beyond Earth. In this area, students explore the coming of the universe story by studying the solar system and stars. They learn the order and characteristics of the Sun, planets, moons, and distant stars, which gives them a sense of scale and place in the universe. This work arises naturally after the First Great Lesson and inspires wonder about cosmic origins.

Curriculum Area: Science (Cosmic Education)
Subject / Strand: Astronomy
Lesson Title: *Exploring the Solar System*
Age Range: 6–9 years (Lower Elementary)
Prerequisites / Previous Work: First Great Lesson (Coming of the Universe), basic sensorial concepts of size and distance
Materials: Globes or planet models, solar system chart or mat, constellation maps, planet puzzles, constellation cards, Sun and

star models, planet research cards
Presentation: The Guide retells the birth of the Sun and planets, using models and lights to show orbits and rotations. Constellations are introduced with charts or lights in a darkened room.
Language: planet, orbit, galaxy, constellation, comet, asteroid, Milky Way, telescope, phase
Points of Interest: Comparing Earth's size to Jupiter; myths of constellations; observing night skies
Direct Aims: Identify planets; explain orbits; understand day/night and seasons; recognize constellations
Indirect Aims: Develop fine motor control with puzzles; expand vocabulary; foster humility through cosmic perspective
Control of Error: Puzzles and matching cards ensure accuracy
Extensions: Observe the Moon or Jupiter's moons; build scale models; link to biology through habitability
Follow-up Work: Space Journals, oral reports, constellation matching games
Notes for the Guide: Begin concrete, move toward abstraction, encourage collaborative projects

Physics

Physics lessons reveal the forces and energy underlying the physical world. Children investigate gravity, magnetism, light, sound, and heat through simple experiments. These explorations show that natural laws govern both the cosmos and daily life.

Curriculum Area: Science (Physical Science)
Subject / Strand: Physics
Lesson Title: *Forces and Energy in Nature*
Age Range: 6–9 years
Prerequisites: Sensorial work with weights and pressure; basic counting and measuring
Materials: Magnets and iron filings, springs, ramps, prisms, tuning forks, flashlights, scales, weights

Presentation: Demonstrate magnetism with filings, gravity with falling objects, light with prisms, sound with tuning forks. Invite predictions and reflections.
Language: gravity, magnetism, force, inertia, friction, conductor, spectrum, vibration, heat, temperature
Points of Interest: Invisible forces, dramatic demonstrations, real-life applications
Direct Aims: Observe and describe forces; identify simple machines; explain light and sound; measure weight
Indirect Aims: Practice scientific method; enhance fine motor skills; develop concentration
Control of Error: Experiments themselves give feedback; circuits only complete if correct
Extensions: Build circuits; test magnet strength; connect physics to astronomy and Earth science
Follow-up Work: Lab Books, oral explanations, construction challenges
Notes for the Guide: Stress observation over formula; ensure safety; tie concepts back to cosmic origins

Chemistry

Chemistry introduces matter and its transformations. Through observation and experiment, children see solids, liquids, and gases change and combine.

Curriculum Area: Science (Physical Science)
Subject / Strand: Chemistry
Lesson Title: *Matter and Its Changes*
Age Range: 6–9 years
Prerequisites: Sensorial classification, pouring and measuring work
Materials: Water, ice, salt, oil, vinegar, baking soda, beakers, filters, natural materials for extraction
Presentation: Show states of matter, dissolving, mixing, and simple

reactions. Use clear, simple vocabulary.
Language: solid, liquid, gas, dissolve, mixture, atom, molecule, reaction, evaporation, condensation
Points of Interest: Reactions like fizzing; crystals under magnification; links to cooking and nature
Direct Aims: Differentiate states of matter; observe mixtures and reactions; introduce elements
Indirect Aims: Build sequential thinking, precision, and patience
Control of Error: Experiments confirm themselves; classification cards self-correct
Extensions: Play with pH indicators, grow crystals, explore environmental chemistry
Follow-up Work: Exploration sheets, oral questioning, card sorting
Notes for the Guide: Emphasize observation, relate to everyday life, allow repetition

Geology

Geology anchors the cosmic story in the Earth itself. Children classify rocks, build models of volcanoes, and simulate erosion, learning that the Earth carries its history in stone.

Curriculum Area: Science (Earth Science)
Subject / Strand: Geology
Lesson Title: *Earth Materials and Landforms*
Age Range: 6–9 years
Prerequisites: Continent/ocean charts, sensorial texture work, outdoor exploration
Materials: Rock/mineral samples, magnifiers, layered sand models, erosion tanks, fossils, tectonic puzzles
Presentation: Show stratified layers, erupt a model volcano, simulate erosion with sand and water
Language: igneous, sedimentary, metamorphic, erosion, volcano, earthquake, fossil, crystal, mineral
Points of Interest: Volcanoes, fossils, shiny minerals, local

geological features
Direct Aims: Classify rocks, name landforms, explain basic plate tectonics, observe earth processes
Indirect Aims: Sharpen observation and documentation, foster environmental awareness
Control of Error: Sorting categories and puzzles self-correct
Extensions: Create geologic timelines, collect rocks outdoors, study fossils and minerals in detail
Follow-up Work: Rock Journals, island-building projects, oral definitions
Notes for the Guide: Connect geology to math, history, art, and environmental renewal

Geography

Geography situates the child on Earth, among continents, oceans, climates, and cultures.

Curriculum Area: Cultural Studies (Geography)
Subject / Strand: Geography
Lesson Title: *The World and Its Peoples*
Age Range: 6–9 years
Prerequisites: Land/water form work, continent charts, compass directions
Materials: Globes, continent puzzles, maps, flags, climate charts, compasses
Presentation: Use the globe to show hemispheres, latitude/longitude, and climates. Puzzles and stories deepen cultural connections.
Language: continent, country, ocean, latitude, longitude, equator, climate, biome, culture
Points of Interest: Locating home, hearing stories of distant peoples, seasonal comparisons
Direct Aims: Identify continents and oceans, read maps, understand climates

Indirect Aims: Build spatial awareness, foster empathy, support research skills
Control of Error: Puzzles and maps provide built-in correction
Extensions: Field trips, neighborhood mapping, travel guides, weather diaries
Follow-up Work: Country reports, quizzes, surveys, oral recall
Notes for the Guide: Be inclusive and culturally sensitive, always tie back to cosmic interconnectedness

Measurement of Time

Measurement of time links the cosmic story to the human story. It shows how Earth's rhythms shape human calendars and history.

Curriculum Area: Cultural Studies (Chronology)
Subject / Strand: Measurement of Time
Lesson Title: *Marking Time: Calendars and Timelines*
Age Range: 6–9 years
Prerequisites: Counting, sequencing, Great Lesson's timeline perspective
Materials: Clocks, calendars, timers, sundials, personal and cosmic timelines
Presentation: Teach time-telling, calendar cycles, and long historical timelines. Contrast personal age with cosmic ages.
Language: second, minute, hour, day, month, year, decade, century, BCE/CE, leap year
Points of Interest: Personal birthdays, vast cosmic time vs short human spans, ancient clocks
Direct Aims: Tell time, use calendars, sequence events, grasp historical eras
Indirect Aims: Strengthen math, planning, and perspective-taking
Control of Error: Timelines and manipulatives self-correct by logical sequence
Extensions: Explore ancient calendars, time zones, moon phases
Follow-up Work: Personal time capsules, quizzes, ordering cards

Notes for the Guide: Tie back to cosmic story, keep concrete, use classroom routines to anchor

Together these six studies give form to the First Great Lesson's invitation. They are not separate subjects but interwoven paths, each reinforcing the others. The child learns that the world is ordered, lawful, beautiful, and deeply connected.

And it is from this foundation that we now step forward. The Great Lessons situate the child within the vast unfolding of the universe, awakening awe for origins and belonging within the great arc of time. Yet Montessori reminds us that just as the cosmos evolves in ordered stages, so too does the human being. To guide children in a world reshaped by artificial intelligence, automation, and social transformation, we must not only tell the story of the universe but also understand the story of human growth itself.

Chapter 13

The Four Planes of Development

It is from this foundation that we now step forward. The Great Lessons situate the child within the vast unfolding of the universe, awakening awe for origins and belonging within the great arc of time. Yet Montessori reminds us that just as the cosmos evolves in ordered stages, so too does the human being. To guide children in a world reshaped by artificial intelligence, automation, and social transformation, we must not only tell the story of the universe but also understand the story of human growth itself.

Dr. Maria Montessori envisioned this unfolding as a rhythmic progression through four distinct planes of development, each lasting approximately six years. These planes are not linear steps but alternating waves of creative transformation and steady consolidation. The first and third planes are highly formative, imaginative, and transformative, while the second and fourth are calmer periods of integration and refinement. Each plane is marked by sensitive periods in which the child is especially receptive to particular experiences, and each carries its own guiding question and developmental task. Taken together, the four planes trace the construction of the human being, mind, body, character, and spirit, from unconscious beginnings to conscious moral responsibility.
In his 1996 article in *The NAMTA Journal*, Camillo Grazzini offered two diagrammatic interpretations of this developmental schema, highlighting the rhythm of alternation between creation and consolidation.

We begin with the **First plane: infancy (0–6 years)**. This is the age of the absorbent mind, when the child takes in the world effortlessly, first unconsciously in the years from birth to three, and then with conscious awareness from three to six. The guiding question of this plane is simple yet profound: ***"What is it?"*** It is the period when order is sought, language is acquired, movement is coordinated, and the senses are refined. Here the foundations of independence are laid, physical, emotional, and cognitive, through real work in the environment. Montessori called this stage the most creative in human life, for it is the time when the personality itself is constructed. By the end of the plane, the child emerges not only with the capacity for self-care and expression, but also with the first integration of the conscious personality, ready to step outward into the wider world.

The **Second plane: childhood (6–12 years)** opens with a different rhythm. Now the child, no longer content with absorbing what is, begins to ask ***"Why?"*** This is a calmer period, one of steady expansion and integration. The reasoning mind blossoms, fueled by imagination and a hunger for the vast story of the universe and humanity's place within it. Sensitive to fairness and justice, children in this stage are deeply interested in moral order and the rules that govern community life. They thrive in groups, working collaboratively, debating questions of right and wrong, and stretching their intellect to discover patterns, laws, and connections. Through Cosmic Education, they are offered the panorama of knowledge, from the birth of stars to the history of civilizations, so they can situate themselves in a world that is both immense and interconnected. The outcome of this plane is intellectual independence, accompanied by the beginnings of moral consciousness.

With the **Third plane: adolescence (12–18 years)**, a new birth takes place. Just as in infancy, the young person must reconstruct themselves, but now as an adult-in-the-making. The guiding question becomes ***"How do I fit in?"*** Adolescents seek dignity, fairness, and belonging, while grappling with fragility, emotional intensity, and the uncertainties of identity. They need protection and encouragement, yet they also long for real responsibility, meaningful work, and opportunities to prove themselves. Sensitive to justice and social participation, they are naturally drawn to questions of ethics, community, and contribution. Montessori envisioned the adolescent in a setting that combines intellectual exploration with real labor, such as her Erdkinder model, where students could cultivate the land, run enterprises, and govern themselves. Through such experiences they gain valorization: the profound sense of worth that arises when their contributions are genuinely needed. The outcome of this plane is an emerging adult identity, grounded in responsibility and prepared for active participation in society.

Finally, the **Fourth plane: maturity (18–24 years)** brings consolidation and direction. Here the questions deepen: ***"Why am I here?"* and *"How far can I go?"*** The young adult, having passed through the turbulence of adolescence, now seeks spiritual and moral independence, discerning a personal vocation and committing to values that will guide life's work. This is a calmer stage, one of refinement and integration, as the individual assumes responsibility for themselves and for their place in the larger human community. Sensitive to conscience and virtue, the young person is now capable of making choices rooted not only in ambition but in contribution to the common good. The outcome of this plane is a complete and integrated adult, equipped to live with moral clarity, to work with integrity, and to offer their gifts in service of humanity.

Digital Citizenship Across the Four Planes

In Montessori's time, the great cultural tools were language, mathematics, and the simple machines of industry. Today, the cultural landscape includes digital networks and artificial intelligence. To remain faithful to Montessori's vision, digital citizenship cannot be treated as a separate subject. It may be received as part of the child's cultural inheritance, integrated gently, age-appropriately, and always subordinated to the deeper work of self-construction.

Infancy (0–6 years). At this plane, digital citizenship is not about direct use of devices but about orientation, order, and grace with tools in the environment. The child absorbs the behaviors of adults: how we handle objects, how we speak to one another, and how we interact with our surroundings. When adults model politeness with digital tools, putting them away after use, naming them accurately, treating them as helpers rather than distractions, the child unconsciously normalizes the idea that technology exists but does not dominate. Here the emphasis is not digital skill but the cultivation of independence, order, and respect for the environment.

Childhood (6–12 years). With imagination and reasoning awakened, the child hungers for explanation and fairness. Digital citizenship enters naturally as part of *responsible research*: learning how to evaluate information, respect authorship, and recognize the difference between truth and error. Guided digital searches, lessons in source evaluation, and early explorations of coding through concrete games mirror the child's growing ability to think abstractly. Because this age group is highly sensitive to justice, discussions about honesty online and fairness in digital spaces become extensions of their natural moral development.

Adolescence (12–18 years). This is the age of identity, dignity, and contribution. Digital citizenship now becomes part of *real work*. Adolescents use digital tools for enterprises, community initiatives, and collaborative scholarship. They wrestle with issues of privacy, bias, and equity, guided by their moral sensitivity to fairness and their yearning to belong meaningfully in society.

Here, digital projects are not abstract lessons but genuine contributions, producing media, building websites, or organizing digital campaigns that serve a community. Valorization arises when their digital work is seen, needed, and valued.

Maturity (18–24 years). In the final plane, digital citizenship merges with moral and spiritual independence. The young adult now discerns not just how to use technology, but when to refrain from it. Questions of vocation and conscience arise: What role should technology play in my life's work? How do I safeguard information, respect human dignity, and align professional choices with personal values? At this stage, digital citizenship is no longer taught from outside but lived from within, as part of the integrated adult personality.

As we have seen, the Four Planes of Development provide a compass for understanding how human beings construct themselves across successive stages of life, each with its own sensitivities, questions, and tasks. To meet the realities of our own time, we now turn to the work of mapping: how this developmental framework can guide us in preparing children not only for the eternal tasks of human growth, but also for the cultural tools, technologies, and systems that shape the world they are inheriting.

Chapter 14
Why This Mapping Exercise?

The world is undergoing a profound transformation driven by artificial intelligence, automation, and emerging digital technologies. In the Montessori community, this awareness has been carried forward by the eleven members of the American Montessori Society Innovation Fellowship Think Tank (March 2025): Debbie Adams, Dana Anderson, Pat Barton, Kate Broughton, Sam Gleisten, Aishwarya Saminathin, Shelley Smith, Ronke Okesola, Katie Torres, Michelle Velho, and Mara Weitzman. Each contributed to envisioning how Montessori environments can be intentionally designed for a future shaped by technological, social, and environmental transformation. Among these projects, Aishwarya Saminathan's *The Coming of Technology*, developed as part of the Innovation Labs: The Future Montessori Classroom, stands as a vivid example of how the arc of the Great Lessons can be extended into our present technological era. This collective work, shared in March 2025, asks not only how we might prepare children for the tools they will encounter, but also how we might prepare them to question, to create, and to act with integrity within the systems they will inherit. This is why the mapping becomes essential: it allows us to hold Montessori's developmental vision in one hand and the demands of our technological age in the other, designing environments that prepare children for both.

The Fourth Industrial Revolution, a term coined by Klaus Schwab in his 2016 book *The Fourth Industrial Revolution*, describes an era where advanced technologies, artificial intelligence, biotechnology, the Internet of Things (IoT), nanotechnology, robotics, and quantum computing, blur the boundaries between the physical, digital, and biological worlds. These innovations are reshaping societies and redefining the skills required to thrive in the workforce. Far from requiring an entirely new educational approach, this book highlights how the Montessori curriculum **already** nurtures these competencies through its unique pedagogical design.

This mapping process demonstrates that Montessori pedagogy already provides a comprehensive framework for developing the essential competencies needed in the Fourth Industrial Revolution and beyond. Dr. Maria Montessori's visionary approach has long emphasized the ability to adapt, think critically, and engage meaningfully with the world, qualities that are now more relevant than ever.

This mapping exercise serves as both a research-driven guide and a call to action for Montessori educators, leaders, and advocates. Rather than introducing new concepts, it reinforces the breadth and depth of the Montessori curriculum as it exists today. The foundational skills necessary for future workplaces, many of which have yet to be imagined, are already embedded within Montessori environments. By articulating these connections clearly, we can engage stakeholders, including educators, parents, and policymakers, in meaningful conversations about the role of Montessori education in shaping the future.

Montessori education exists in a unique tension between tradition and technological progress. Historically, Montessori guides have prioritized hands-on, real-world experiences, aligning with Dr. Montessori's belief that children learn best through active engagement with their environment. However, in an era where technology is redefining human interaction, industries, and education itself, there is a pressing need to bridge the gap between Montessori philosophy and the realities of a digital world.

The Montessori community, while deeply insightful and forward-thinking, has often struggled to communicate its relevance in broader educational discourse. As a result, the significant contributions of Montessori pedagogy, particularly its ability to prepare children for an AI-driven society, have been overlooked in mainstream discussions on future-ready education. This book seeks to change that by providing a clear framework that connects Montessori learning with AI competencies and digital-age skills.

Key Realities Addressed in This Book

- **Montessori Education is Not Isolated**
  Montessori schools do not exist in a vacuum; they are part of the broader social and technological landscape. Parents working in AI-driven industries seek an education that equips their children with digital wisdom, not just digital access. They want their children to develop creativity, critical thinking, and deep engagement with the real world, qualities that Montessori already fosters.

- **Technology and AI are Reshaping the Future**
  The world is experiencing unprecedented shifts in the workforce, education, and human interactions due to AI. The Montessori method, with its emphasis on observation, adaptation, and self-directed learning, is uniquely positioned to prepare children not just to use technology, but to understand and shape it responsibly.

- **The Need for a Shared Vision and Research-Based Framework**
  Montessori education has always been rooted in scientific observation and research. This book documents and maps Montessori materials to AI competencies, ensuring that Montessori educators have an initial structure for guiding children in the digital age. By providing a framework for

exploration, we empower schools, guides, and families to confidently articulate how Montessori aligns with the skills needed for the future.

- **Montessori Has Always Adapted to the Times**
  Dr. Maria Montessori developed her method in response to social, political, and scientific changes. She worked through times of war, revolution, and educational transformation, always adapting to meet the needs of the moment.
  In that same spirit, this book recognizes that while Montessori materials remain consistent, the world in which they exist has changed. We must actively demonstrate how Montessori education naturally aligns with the competencies required for the future and ensure that educators, parents, and schools recognize this alignment.

- **A Thoughtful, Intentional Approach to AI**
  Unlike educational approaches that may integrate technology without considering its deeper implications, Montessori education moves with care and purpose. This book does not advocate for unchecked AI integration but rather highlights how Montessori's strengths, problem-solving, interdisciplinary thinking, and hands-on engagement, already prepare children for an AI-driven world.

To ground this mapping in practical reality, we now turn to the materials themselves. The following chapter presents selected examples of Montessori materials across developmental stages, examining their original purpose and revealing how they naturally cultivate the competencies needed for an emerging technological era.

Chapter 15

Montessori Materials & Their Purpose

Montessori materials are not the guarded property of any one organization, nor are they confined to a single interpretation. They are the physical embodiment of an approach to learning that trusts the child's mind, hands, and heart to work in harmony. Each material holds within it an idea, a discovery, a relationship to the world. They are, in essence, keys, opening doors to mathematics, language, science, and art, while simultaneously cultivating habits of attention, perseverance, and wonder.

In the pages that follow, I offer general descriptions that are broad in scope and accessible in spirit. They are not bound to the prescriptive language of any handbook, but instead aim to give all educators, parents, and thinkers a clear sense of what these materials are and why they matter. This is an invitation to see them as they are: timeless tools for human development, ready to meet the needs of children in every community, in every era, including our own age of rapid technological change.

Infant & Toddler (Ages 0-3)

Practical Life Materials

- Pull-Up Bars and Mirrors – Support balance, coordination, and self-awareness.
- Low Shelves with Accessible Materials – Encourage independence and decision-making.
- Child-Sized Furniture – Foster self-reliance in daily activities.
- Self-Care Stations – Promote personal hygiene and autonomy.
- Grasping Toys & Rattles – Develop fine motor skills and hand-eye coordination.

Sensorial Materials

- Object Permanence Boxes – Strengthen memory and understanding of object continuity.
- Nesting and Stacking Materials – Enhance spatial awareness and coordination.
- Texture Boards and Fabric Swatches – Refine tactile discrimination.
- Sound Cylinders – Develop auditory differentiation.
- Treasure Baskets – Stimulate sensory exploration with natural objects.

Language Materials

- Picture Books with Realistic Images – Build vocabulary and comprehension.

- Language Cards and Classified Objects – Introduce categorization and early naming.
- Simple Musical Instruments – Encourage sound recognition and rhythm.

Motor Skill Materials

- Bead Stringing Sets – Develop fine motor control and concentration.
- Peg Boards and Puzzles – Strengthen problem-solving and dexterity.
- Rolling Pins and Dough – Encourage creativity and hand strength.

Cultural Materials

- Multicultural Dolls and Figurines – Foster diversity awareness and emotional connections.
- Nature Exploration Tools – Introduce outdoor experiences and natural elements.

Early Childhood (Ages 3-6)

Practical Life Materials

- Pouring Exercises – Develop fine motor skills, independence, and focus.
- Spooning and Transferring Activities – Enhance coordination and writing readiness.
- Dressing Frames – Teach self-care skills such as buttoning and zipping.
- Cleaning Tools – Encourage responsibility and care for the environment.

Sensorial Materials

- Pink Tower – Introduce size gradation, spatial awareness, and early math concepts.
- Brown Stair – Support spatial awareness and visual perception.
- Red Rods – Develop length discrimination and mathematical relationships.
- Knobbed Cylinders – Enhance visual and tactile perception.
- Color Tablets – Teach color recognition and matching.

Language Materials

- Sandpaper Letters – Facilitate tactile and phonetic learning of letters.
- Moveable Alphabet – Support word construction and early writing.

- Metal Insets – Prepare hand control for writing through geometric tracing.
- Object and Picture Matching Cards – Expand vocabulary and comprehension.

Mathematics Materials

- Number Rods – Teach number sequencing and quantity relationships.
- Spindle Box – Reinforce counting skills and the concept of zero.
- Golden Beads – Introduce decimal system and place value.
- Stamp Game – Provide a concrete understanding of arithmetic operations.
- Bead Chains – Aid in learning skip counting, squaring, and cubing.

Cultural Materials

- Puzzle Maps – Develop geography skills and spatial awareness.
- Land and Water Forms – Teach fundamental physical geography concepts.
- Timeline of Life – Foster understanding of evolution and history.
- Botany Cabinet – Introduce plant identification and classification.

Elementary (Ages 6-12)

Mathematics Materials

- Golden Bead Materials – Deepen understanding of place value and operations.
- Checkerboard & Bead Frames – Introduce multiplication and division concepts.
- Fraction Insets – Explore fraction equivalency and operations.
- Algebraic Binomial & Trinomial Cubes – Lay foundations for algebraic thinking.

Language & Grammar Materials

- Grammar Symbols – Represent parts of speech visually.
- Sentence Analysis Materials – Develop sentence structure comprehension.
- Reading Classification Cards – Expand vocabulary and comprehension.

Cultural & Science Materials

- First Great Lesson (Creation Story) – Explore the origins of the universe.
- Second Great Lesson (Coming of Life) – Introduce evolution and biology.
- Third Great Lesson (Coming of Humans) – Learn about anthropology and civilizations.
- Fourth Great Lesson History of Writing) – Investigate the

development of communication.

- Fifth Great Lesson (History of Numbers) – Understand the origins of mathematical thought.

Adolescence (Ages 12-18)

Mathematics & Logic

- Advanced Algebra & Geometry Materials – Support abstract mathematical thinking.
- Business Math & Economics Materials – Teach financial literacy and economic principles.

Language & Humanities

- Debate & Public Speaking Activities – Develop communication and argumentation skills.
- Research & Thesis Writing Materials – Prepare students for academic and real-world writing.

Science &Innovation

- Hands-on Experiments & Lab Work – Reinforce scientific inquiry and exploration.
- Technology & Engineering Materials – Support coding, robotics, and practical STEM applications.

Cultural & Societal Materials

- Micro-Economy & Entrepreneurship Materials – Encourage real-world business skills.
- Global Citizenship & Ethics Discussions – Foster critical thinking about societal issues.
- Community-Based Learning & Internships – Provide experiential education beyond the classroom.

This progression of Montessori materials aligns with Dr. Maria Montessori's philosophy of following the child's development through hands-on, experiential learning. Each material is scaffolded to build upon previous knowledge, fostering independence, critical thinking, and real-world application at every stage of education. To ensure this work remains grounded in both Montessori tradition and contemporary relevance, the following chapter outlines the key sources that informed this mapping, drawing from educational research, global AI competency frameworks, and the foundational writings of Dr. Montessori herself.

Chapter 16

Literature Review

The following documents were reviewed for this research and will form the foundations of further study:

#1 – McKinsey & Company: "Defining the Skills Citizens Will Need in the Future World of Work" (June 2021)

#2 – UNESCO: AI competency framework for Students (2024)

#3 – World Economic Forum: "Schools of the Future: Defining New Models of Education for the Fourth Industrial Revolution" (January 2020)

The following information has been selected for its competency content. Extracted images and information have been labeled with the title and page numbers from the original documents.

Resource #1

McKinsey & Company: "Defining the Skills Citizens Will Need in the Future World of Work" (June 2021)

Exhibit 1 - Our research identified 56 foundational skills that will help citizens thrive in the future of work. Our research identified 56 foundational skills that will help citizens thrive in the future of work - Page 3

These 56 DELTAs are grouped into:

4 overarching categories

13 skill groups

56 individual skills, each referred to as a DELTA because they are a combination of:

- A skill (can be learned)
- An attitude (can be nurtured)
- A behavior (can be observed and practiced)

Appendix: Definitions and Methodology (pp. 12-17)

Skill Groups and Elements

Communication

- Active listening
- Asking the right questions
- Storytelling and public speaking
- Synthesizing messages

Critical thinking

- Logical reasoning
- Seeking relevant information
- Structured problem solving
- Understanding biases

Developing relationships

- Empathy
- Humility
- Inspiring trust
- Sociability

Digital fluency and citizenship

- Digital collaboration
- Digital ethics
- Digital learning
- Digital literacy

Entrepreneurship

- Breaking orthodoxies
- Courage and risk-taking
- Driving change and innovation
- Energy, passion, and optimism

Goals achievement

- Achievement orientation
- Coping with uncertainty
- Grit and persistence
- Ownership and decisiveness

- Self-development

Mental flexibility

- Ability to learn
- Adaptability
- Adopting a different perspective
- Creativity and imagination
- Translating knowledge to different contexts

Mobilizing systems

- Crafting an inspiring vision
- Organizational awareness
- Role modeling
- Win–win negotiation

Planning and ways of working

- Agile thinking
- Time management and prioritization
- Work-plan development

Self-awareness and self-management

- Integrity
- Self-confidence
- Self-control and regulation

- Self-motivation and wellness
- Understanding own emotions and triggers
- Understanding own strengths

Software use and development

- Computational and algorithmic thinking
- Data analysis and statistics
- Programming literacy

Teamwork effectiveness

- Coaching
- Collaboration
- Empowering
- Fostering inclusiveness
- Motivating different personalities
- Resolving conflict

Understanding digital systems

- Cybersecurity literacy
- Data literacy
- Smart systems
- Tech translation and enablement

Resource #2

UNESCO. (2024). *Competency framework for students* (F. Miao & K. Shiohira, Authors). UNESCO.

Table 1. AI competency framework for students

Competency aspects	Progression levels		
	Understand	Apply	Create
• Human-centred mindset	• Human agency	• Human accountability	• Citizenship in the era of AI
• Ethics of AI	• Embodied ethics	• Safe and responsible use	• Ethics by design
• AI techniques and applications	• AI foundations	• Application skills	• Creating AI tools
• AI system design	• Problem scoping	• Architecture design	• Iteration and feedback loops

Table 1. AI competency framework for students - Page 19

Human-centred mindset

Competency aspects	Progression levels		
	Understand	Apply	Create
• Human-centred mindset	• Human agency	• Human accountability	• Citizenship in the era of AI

Human-centred mindset - Page 22

Ethics of AI

Competency aspects	Progression levels		
	Understand	Apply	Create
• Ethics of AI	• Embodied ethics	• Safe and responsible use	• Ethics by design

Ethics of AI - Page 23

AI techniques and applications

Competency aspects	Progression levels		
	Understand	**Apply**	**Create**
• AI techniques and applications	• AI foundations	• Application skills	• Creating AI tools

AI techniques and applications - Page 24

AI system design

Competency aspects	Progression levels		
	Understand	**Apply**	**Create**
• AI system design	• Problem scoping	• Architecture design	• Iteration and feedback loops

AI system design - Page 25

Table 1 – Content (built-in mechanisms for skills adaptation)

Global citizenship skills

To include content that focuses on building awareness about the wider world, sustainability, and playing an active role in the global community.

Innovation and creativity skills

To include content that fosters skills required for innovation, including complex problem-solving, analytical thinking, creativity, and systems-analysis.

Technology skills

To include content that is based on developing digital skills, including programming, digital responsibility, and the use of technology.

Interpersonal skills

To include content that focuses on interpersonal emotional intelligence (i.e., empathy, cooperation, negotiation, leadership, and social awareness).

Personalized and self-paced learning
From a system where learning is standardized to one based on the diverse individual needs of each learner, and flexible enough to enable each learner to progress at their own pace.

Accessible and inclusive learning
From a system where learning is confined to those with access to school buildings to one in which everyone has access to learning and is therefore inclusive.

Problem-based and collaborative learning
From process-based to project and problem-based content delivery, requiring peer collaboration and more closely mirroring the future of work.

Lifelong and student-driven learning
From a system where learning and skilling decrease over one's lifespan to one where everyone continuously improves on existing skills and acquires new ones based on their individual needs.

Competency aspects and Progression levels

Page 22 – Human-centred mindset:
Understand: Human agency
Apply: Human accountability
Create: Citizenship in the era of AI

Page 23 – Ethics of AI:
Understand: Embodied ethics
Apply: Safe and responsible use
Create: Ethics by design

Page 24 – AI techniques and applications:
Understand: AI foundations
Apply: Application skills
Create: Creating AI tools

Page 25 – AI system design:
Understand: Problem scoping
Apply: Architecture design
Create: Iteration and feedback loops

Resource #3

World Economic Forum: "Schools of the Future: Defining New Models of Education for the Fourth Industrial Revolution" (January 2020)

Figure 2: The World Economic Forum Education 4.0 Framework

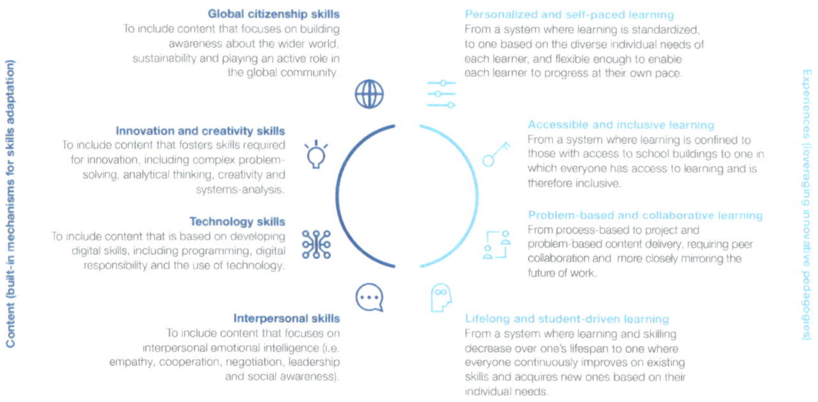

Figure 2: The World Economic Forum Education 4.0 Framework - Page 7

Figure 2. Content:

Content (Built-in Mechanisms for Skills Adaptation)

- Global Citizenship Skills – Focuses on building awareness about the wider world, sustainability, and playing an active role in the global community.
- Innovation and Creativity Skills – Encourages skills required for innovation, including complex problem-solving, analytical thinking, creativity, and systems analysis.
- Technology Skills – Based on developing digital skills, including programming, digital responsibility, and the use

of technology.

- Interpersonal Skills – Focuses on interpersonal emotional intelligence, including empathy, cooperation, negotiation, leadership, and social awareness.

Experiences (Leveraging Innovative Pedagogies)

- Personalized and Self-Paced Learning – Transitioning from standardized learning to a learner-centered system based on individual needs, allowing flexibility in progression.

- Accessible and Inclusive Learning – Expanding education beyond physical schools to ensure universal access, making learning inclusive.

- Problem-Based and Collaborative Learning – Shifting from process-based to project-based and problem-based learning, requiring peer collaboration, aligning with the future of work.

- Lifelong and Student-Driven Learning – Moving from static learning models to continuous skill improvement throughout life, allowing learners to acquire new skills based on evolving needs.

Summary of Foundational Resources for AI Competency Mapping

Together, these three globally recognized frameworks outline a shared vision of the skills and competencies essential for thriving in a world increasingly shaped by artificial intelligence. Each framework contributes a unique perspective while also overlapping in important ways, creating a strong foundation for aligning Montessori education with future-facing competencies.

1. McKinsey & Company: "Defining the Skills Citizens Will Need in the Future World of Work" (2021)

McKinsey introduces a detailed framework of 56 foundational skills, known as **DELTAs** (Distinct Elements of Talent), which represent a mix of attitudes, skills, and behaviors that can help citizens thrive in the future of work. Organized across four overarching categories and thirteen skill groups, the framework provides a comprehensive vision of human and digital competencies. The categories include:

- Cognitive: Critical thinking, communication, mental flexibility, planning and ways of working

- Interpersonal: Teamwork, empathy, leadership, collaboration

- Self-leadership: Self-awareness, resilience, entrepreneurship, goal achievement

- Digital: Digital fluency, programming, data literacy, algorithmic thinking

These competencies are clearly defined in McKinsey's appendix (pp. 12–17), providing a vocabulary and reference point for mapping against Montessori methods.

1. UNESCO: AI Competency Framework for Students (2024)

UNESCO's framework outlines four core competency domains critical for student development in an AI-enabled world. These are:

- Human-centred mindset
- Ethics of AI
- AI techniques and applications
- AI system design

Each competency progresses through three levels: Understand, Apply, and Create. This progression helps educators scaffold AI literacy development across age groups:

- Understand: foundational knowledge and awareness
- Apply: practical usage and accountability
- Create: innovation, design, and ethical AI development

This structure resonates strongly with Montessori's developmental approach and can serve as a roadmap for age-appropriate integration of AI concepts.

1. World Economic Forum: "Schools of the Future: Defining New Models of Education for the Fourth Industrial Revolution" (2020)

The WEF's Education 4.0 framework complements the above by emphasizing both content and pedagogy. Content areas include:

- Global citizenship
- Innovation and creativity
- Interpersonal skills

- Technology skills

Pedagogical principles are also reimagined, with a shift toward:

- Personalized and self-paced learning
- Accessible and inclusive learning
- Problem-based and collaborative learning
- Lifelong and student-driven learning

This dual focus on what students learn and how they learn offers essential guidance for educators preparing children for a future where adaptability, creativity, and ethical reasoning are paramount.

Together, these three frameworks form the evidence base for mapping AI competencies to Montessori pedagogy. They provide a shared language for describing future-ready skills while respecting the autonomy and dignity at the heart of Montessori education. Synthesizing this body of work has been both humbling and challenging, an exercise in deep observation, reflection, and discernment. As with all Montessori endeavors, this is an initial iteration, a thoughtful first step. It is offered not as a finished product, but as an invitation to the wider community, to collaborate, refine, and expand upon it with grace, rigor, and fidelity to the child.

Chapter 17
Synthesizing Montessori & AI Competency Frameworks

Across diverse yet converging frameworks from McKinsey & Company, the World Economic Forum, and UNESCO, a unified message emerges: the future of education and work demands a reimagined skillset. These institutions emphasize that preparing learners for a rapidly evolving, AI-infused world requires a foundation that spans cognitive agility, emotional intelligence, ethical leadership, and technological fluency.

The Foundation of the Work

McKinsey & Company identifies fifty-six foundational skills, called DELTAs, across four broad categories: Cognitive, Interpersonal, Self-leadership, and Digital. These are not abstract traits, but tangible, observable capabilities such as logical reasoning, adaptability, empathy, and programming literacy.

The World Economic Forum's Education 4.0 Framework echoes this need by highlighting pedagogical shifts, from standardized to personalized learning, from passive to collaborative and problem-based approaches. It calls for content that fosters innovation, creativity, technology fluency, interpersonal skills, and global citizenship.

UNESCO takes the conversation further by contextualizing these competencies in relation to artificial intelligence. Their framework outlines progression levels, from understanding, to applying, to creating, across key areas such as human-centered mindsets, ethics of AI, and system design.

Making Montessori the Bridge

As a researcher rooted in the Montessori community, I approached these frameworks not only as roadmaps for future skills, but as mirrors reflecting what is already present in Montessori pedagogy. Dr. Maria Montessori's method, developed more than a century ago, emphasizes hands-on learning, intrinsic motivation, and deep observation of the child, elements that align remarkably with these modern global competencies.

Through careful analysis and collaborative work with AI tools, I mapped the competencies identified by McKinsey, WEF, and UNESCO against the developmental sequences and materials of the Montessori Method. The result is not a reinvention, but a revelation: many of these future-facing skills are embedded in Montessori from birth through adolescence.

How the Competency Table Was Formed

The final table of Montessori-aligned AI competencies was created through a reflective synthesis of the global frameworks and Montessori's developmental approach. Starting with McKinsey's DELTAs and WEF's thematic pillars, I examined how these aligned with Montessori's existing materials and practices by age group. I then layered in UNESCO's AI competency progression, from understanding to applying to creating, to identify where Montessori already supports foundational AI skills, and where there is space to deepen or extend learning.

The resulting list shows which AI competencies are present and integrated at each stage of the Montessori journey: 0–3, 3–6, 6–12, and 12–18. It reflects not only the competencies themselves, but their emergence and maturation across developmental planes.

This synthesis is not a prescription for change, but a recognition of readiness. The Montessori method, with its emphasis on autonomy, practical life, and interconnectedness, already lays the groundwork for learners to thrive in the Age of AI. The table created is a starting point, a call to Montessori educators to see, name, and celebrate the competencies they already foster, and to step forward as leaders in this global educational transformation.

This is our moment to reclaim the narrative: Montessori is not behind the curve of innovation. It is, and always has been, the curve.

Age-Structured AI Competency Progression

Early Exploration & Sensory AI Foundations (Ages 0–3)

Zero Screens

At this stage, learning is rooted in sensory experiences and cause-effect reasoning:

- **Social Intelligence & Empathy** – Developing early awareness of people and emotions.
- **Order & Sequence Awareness** – Recognizing patterns in daily routines and activities.
- **Spatial Awareness** – Understanding object relationships in space.
- **Pattern Recognition** – Identifying shapes, colors, and sequences.
- **Cause & Effect** – Observing how actions lead to outcomes.
- **Hand-Eye Coordination** – Engaging in physical interactions to refine motor skills.
- **Tactile & Sensory Data Processing** – Exploring through touch, sight, and sound.
- **Simple Problem-solving Steps** – Beginning to anticipate and solve basic challenges.

This stage builds the foundation for computational thinking, as children begin to recognize patterns and relationships that later support AI-related cognition.

Introduction to Computational Thinking & Digital Awareness (Ages 3–6)

Zero Screens

At this stage, children develop pre-logical reasoning and structured thinking:

- **Order & Symmetry** – Recognizing balance, patterns, and organization.
- **Motor Skills & Coordination** – Refining physical control through hands-on tasks.
- **Algorithmic Thinking Foundations** – Understanding sequences and simple step-by-step processes.
- **Observational Skills** – Developing attention to patterns and logical structures.
- **Data Sorting & Classification** – Grouping objects based on similarities and differences.
- **Attention to Detail** – Enhancing precision in identifying and categorizing information.
- **Digital Citizenship as Practical Life** – Learning early technology etiquette and digital responsibility.

This stage introduces AI ethics in simple terms, ensuring young learners develop a foundational understanding of technology's role in their environment.

Developing AI Awareness & Ethical Thinking (Ages 6–12)

Early Screen Use as Practical Life

In this phase, children refine logical reasoning and problem-solving skills:

- **Precision & Control** – Developing accuracy in decision-making and digital interactions.
- **Sequential Logic** – Understanding step-by-step processes in problem-solving.
- **Basic Computational Thinking** – Learning structured problem-solving approaches.
- **Conceptual Understanding of Patterns** – Identifying trends and making predictions.
- **Basic Data Analysis** – Interpreting and organizing information for insights.
- **Problem-solving with Complexity** – Engaging in multi-step challenges.
- **Algorithm Creation** – Designing logical sequences for tasks.
- **Ethical Considerations** – Beginning discussions on fairness, bias, and responsible AI use.

By embedding AI within meaningful, project-based learning, students see how technology impacts their world while building practical digital and ethical awareness.

Advanced AI Literacy & Human-Centered Design (Ages 12–18)

Discerning Screen Use for Research and Creativity

At this level, students engage in critical thinking, abstract problem-solving, and ethical AI discussions:

- **Mathematical Foundations in AI** – Applying algebra, statistics, and logic to AI concepts.
- **Indirect Preparation for Natural Language Processing** – Understanding language structures and semantics.
- **Advanced Computational Thinking** – Engaging in abstract reasoning and algorithmic problem-solving.
- **Data Structuring & Analysis** – Organizing and interpreting complex datasets.
- **Data Visualization** – Presenting insights through charts, graphs, and models.
- **Algorithm Optimization** – Enhancing efficiency in computational solutions.
- **Machine Learning Foundations** – Exploring how AI learns from data and makes predictions.
- **Project-based AI Applications** – Developing real-world AI solutions.
- **AI Ethics & Self-Reflection** – Evaluating AI's impact on society and personal responsibility in its use.

This stage equips students with the analytical, technical, and ethical decision-making skills needed for AI fluency in higher education and the workforce.

Why This Framework Matters

This approach ensures that AI education is:

- **Cognitively aligned** – Based on developmental psychology (Montessori).
- **Scientifically grounded** – Informed by leading AI competency frameworks (UNESCO, McKinsey, WEF).
- **Pedagogically meaningful** – Embedded in experiential, inquiry-based, and ethical learning.

By integrating AI literacy gradually and thoughtfully into education, grounded in developmental principles and informed by current research, we honor the Montessori tradition of meeting the child where they are. This integration is not a one-size-fits-all prescription but a process that must be continually observed, reflected upon, and adapted. The methodology that underpins this work arises from a deep respect for the individual learner, the integrity of Montessori materials, and the evolving context of the digital age. What follows is a reflective inquiry into how alignment between AI competencies and Montessori pedagogy has been approached, the reasoning behind specific choices, and the process of navigating this emerging landscape with grace and curiosity.

Chapter 18

Methodology Notes: A Reflective Inquiry into Alignment

This work did not begin as a technical mapping exercise. It began, as all Montessori observation does, with a question:

"Is it possible that the competencies the world now urgently needs, ethical reasoning, cognitive agility, systems thinking, are already present in the prepared Montessori environment?"

Rather than constructing a hypothesis and testing it through fixed metrics, as this research was done gradually over a period of two years, I approached this in the "building the plane while flying it" method, and created a qualitative synthesis. I immersed myself in the most widely cited global frameworks identifying future-ready skills: McKinsey & Company's DELTA model, the World Economic Forum's Education 4.0 shifts, and UNESCO's competency progression for AI in education. These frameworks offer clear insight into what today's children will need to navigate an AI-saturated world, skills not only of computation, but of human discernment, self-regulation, and creative, adaptive thinking.

At the same time, I turned inward, to the Montessori method itself. I studied its materials, sequences, and pedagogical assumptions across the planes of development. I drew on the American Montessori Society's suggested materials lists, classroom observations, NCMPS rubrics, and decades of training albums. I read Montessori's own words and those of her interpreters and reformers, listening closely to the voice of a pedagogy grounded in a scientific respect for the unfolding child.

From there, I began the process of alignment, not as a checklist, but as a meditation. Using a grounded theory approach, I allowed connections to emerge: a sensorial activity matched with pattern recognition; the Three Period Lesson with recall development and formative assessment; the adolescent's self-initiated projects with algorithmic thinking and data analysis. I observed not only what competencies were present in the materials, but *when* they emerged in the child's timeline, how they ripened, how they repeated, how they deepened.

The process was iterative. I returned to the tables again and again, testing each alignment not for perfection, but for integrity. Where Montessori seemed to exceed the frameworks, offering not just competence but consciousness, I noted that too. Where gaps appeared, I did not treat them as failures, but as invitations for further inquiry and development.

At times, AI itself became a research companion. Using language models as reflective partners, I tested phrases, questioned assumptions, and refined structure. But the insights remained rooted in human experience, in the classroom, in the child, in the method.

This methodology is not empirical in the traditional sense, but it is no less rigorous. It is built upon layered observation, deep listening, careful synthesis, and reflective practice. It is offered not as a final conclusion, but as a beginning: an open-source invitation for others in the Montessori and educational communities to continue mapping, testing, and evolving.

Because in the end, this is not only a project about competencies. It is a project about confidence. The confidence to see that Montessori may already hold the answers to questions we are only now beginning to ask. With this in mind, the following section offers a developmental mapping of Montessori materials aligned with AI competencies, organized by age, rooted in observation, and grounded in the belief that children, when given the right environment and guidance, are more than capable of growing into ethical, capable, and future-ready citizens. This map is not definitive, but it is a starting point, a living framework to inspire further exploration, adaptation, and shared learning.

As we continue, we recognize that translating vision into practice is where the real work begins. While the developmental mapping offers a broad view of possibilities, many guides, especially those newer to Montessori or to AI, are looking for practical, structured ways to bring this alignment to life in the classroom. How might we move from theory to touchable action, from a framework to daily practice? How do we ensure that children's innate curiosity and independence remain central, even as we introduce new forms of thinking shaped by the age of artificial intelligence?

An inspiring example of early integration of digital literacy within Montessori pedagogy can be seen in the work of Desislava Stoeva, Assistant Professor at Sofia University St. Kliment Ohridski, Faculty of Educational Studies and the Arts. Through action-research in 3–6 classrooms, Stoeva and her team developed a set of Montessori-aligned materials and practices to foster media and digital literacy in early childhood without relying on screens. Her work focuses on six foundational domains: information literacy, communication and collaboration, digital content creation, safety, problem-solving, and critical thinking. These domains are introduced using hands-on, multi-sensory materials that respect Montessori developmental principles. The course, titled *AMS Learning Live: How Do You Teach Media and Digital Literacy in Montessori Preschools?*, is available through the American Montessori Society and provides guides with practical strategies for preparing young digital citizens within the Montessori environment.

Chapter 19

Sample: Extensions & Applications Supporting the New Guide

Montessori Foundations for AI-Relevant Thinking (Unplugged)

This section responds to thoughtful feedback from early reviewers of this work who asked for concrete illustrations to support new and emerging guides as they begin to recognize how AI-relevant competencies already live within authentic Montessori environments.

Before proceeding further into the formal mapping of competencies, frameworks, and research, this chapter offers a small window into what is already possible using time-honored Montessori materials and practices. These examples are not a curriculum, nor a prescribed sequence. They are illustrations, designed to make visible what experienced Montessori educators already know: that abstraction, logic, sequencing, classification, pattern recognition, and inference emerge naturally through hands-on, child-led work.

Importantly, **no step toward digital technology is required**. The activities described here are fully complete as human, material-based learning experiences. In Montessori communities that later choose to explore digital tools, robotics, or AI systems, often through thoughtfully prepared makerspaces, these foundational capacities provide a grounded, ethical, and developmentally aligned bridge. For many communities, the work described here remains sufficient in itself.

Glossaries and terminology referenced here appear at the end of the book for guides who wish to deepen their conceptual language.

Ages 3–6

Practical Life & Sensorial Foundations (Unplugged)

Activity 1: Daily Routine Sequencing

Montessori Domain: Practical Life
AI-Relevant Capacity: Sequencing, precision, logical order

A familiar daily task, such as handwashing or table washing, is observed and performed by the child. The child names each step, refining order and clarity. Optional picture cards may be used to externalize the sequence. A "program a friend" game invites the child to give precise instructions to a guide, highlighting how small ambiguities change outcomes.

Developmental Rationale: At this stage, children are constructing internal order, coordination, and intentional movement. Sequencing emerges through action, not abstraction.

Activity 2: Pink Tower Strategy Description

Montessori Domain: Sensorial
AI-Relevant Capacity: Implicit algorithms, debugging, spatial reasoning

The child builds the Pink Tower in standard form. The guide invites the child to describe how they approached the task. The child then "programs" the guide to recreate the tower using exact language. When errors occur, the child identifies and corrects them. Variations include reverse order or intentional mistakes.

Developmental Rationale: Children aged 3–6 think through movement and material. Describing strategy strengthens metacognition without introducing symbolic code.

Activity 3: Sorting Mystery Objects

Montessori Domain: Sensorial / Language
AI-Relevant Capacity: Classification, rule formation, data grouping

A mixed set of objects is sorted by the child according to a self-chosen rule. The child explains the logic, then resorts using a different feature. A "guess my rule" exchange makes reasoning visible. The guide may gently connect this to how systems learn to group information.

Developmental Rationale: Classification at this age is concrete and relational. The emphasis is on perception and explanation, not correctness.

Activity 4: Sound Cylinder Patterning

Montessori Domain: Sensorial
AI-Relevant Capacity: Pattern recognition, auditory discrimination

The child pairs sound cylinders, then grades them from quietest to loudest. The child describes perceived differences. Real-world connections may be made to how humans and tools recognize sound patterns.

Developmental Rationale: Refinement of the senses is foundational to later abstraction. Pattern recognition begins in the body.

Activity 5: Mystery Bag Inference

Montessori Domain: Sensorial
AI-Relevant Capacity: Inference, feature detection, hypothesis testing

Objects are placed in a mystery bag. The child infers identity through touch, naming features that informed their guess. Roles may be reversed. Optional questioning games introduce early decision-tree logic through dialogue.

Developmental Rationale: Inference develops through sensory feedback and language, not probability models.

Ages 6–9

Elementary Extensions (Still Unplugged)

At the elementary level, the same materials and games evolve through **reflection, comparison, and peer dialogue**. Children may:

- Compare multiple sequences and debate efficiency
- Invent their own classification systems
- Create flowcharts using paper, arrows, and symbols
- Document strategies in journals or on classroom charts

This work remains entirely material-based and collaborative.

Ages 9–12

Prepared Makerspaces as an Optional Bridge

In some Montessori communities, especially at the upper elementary level, **makerspaces emerge as a natural extension of the prepared environment**. These spaces emphasize construction, repair, experimentation, and systems thinking, not screen dependency.

When communities choose to explore programmable tools, robotics, or AI concepts, the transition is not a leap but a continuation. Children approach tools as researchers, not consumers, grounded in years of embodied logic, ethical discussion, and material intelligence.

Participation in such spaces is a **community decision**, not a developmental requirement.

These activities can be rotated throughout the year and adapted to each guide's environment. They offer accessible, Montessori-aligned ways to cultivate foundational AI-relevant capacities, such as sequencing, classification, pattern recognition, and inference, using existing materials and human interaction alone. No digital tools are required. The learning is complete in itself, grounded in movement, sensory feedback, language, and relationship.

Most importantly, these examples affirm what Montessori education has always demonstrated: that the child's natural capacity for abstraction, logic, experimentation, and ethical reasoning emerges through hands-on work in a carefully prepared environment. These capacities are essential not only for understanding artificial intelligence, but for navigating life with clarity, agency, and responsibility.

As educators begin to recognize these natural connections, questions understandably arise. What does this look like in practice for our youngest learners? How might time-honored Montessori materials make visible the same forms of thinking now described in contemporary language as "AI competencies," without rushing children toward screens or systems?

In some communities, particularly at the elementary level, thoughtfully prepared Montessori makerspaces offer one possible extension of this work. When approached as environments for construction, repair, experimentation, and inquiry, rather than technology consumption, makerspaces become a natural continuation of the Montessori prepared environment. They allow children who are developmentally ready to explore tools and systems with discernment, grounded in years of unplugged, material-based reasoning.

Whether a community chooses to take this step or not, the foundation remains the same. The work begins with the child, the materials, and the environment. Technology, when it appears, follows human development, not the other way around.

Chapter 20

Sample: Mapping 3–6 Montessori Materials to AI Competencies

In response to the EC Coordinator in Shenzhen, this chapter offers an example of a structured table mapping Montessori materials for Early Childhood (ages 3–6) to foundational AI competencies. The goal is not to replace the core material or the teacher's role, but to highlight how existing Montessori practice prepares children for AI-relevant thinking. The extensions are meant to support, not replace, the framework.

Montessori Material (Area)	AI Competency	Teacher Observation	Extension Idea
Handwashing Routine (Practical Life)	Algorithmic thinking (sequencing)	Follows steps in order, self-corrects if one is missed	Picture cards, "Program a Friend" game, sequencing verbal commands
Pink Tower (Sensorial)	Ordered reasoning Algorithm design Spatial awareness	Builds tower using a consistent internal rule	Describe strategy, teach someone else step-by-step, reverse-build, debug
Sorting Cards (Language/Culture)	Data classification	Sorts by feature (animal/vehicle), explains logic	"Guess My Rule" game, compare different rules for same set, sort by multiple features
Sound Cylinders (Sensorial)	Pattern recognition (audio)	Matches sound pairs, can grade by volume	Grade from quiet to loud, describe sound types, relate to voice recognition or AI sound detection
Mystery Bag (Sensorial)	Inference, feature recognition	Identifies objects by feel, explains reasoning	Discuss features used to guess, swap roles, 20-Questions-style description game

Sample: Mapping 3–6 Montessori Materials to AI Competencies

By making these connections explicit, this mapping begins to show how Montessori pedagogy already prepares children to meet the demands of an AI-influenced world. The competencies often described as "new" emerge naturally through observation, concrete materials, and hands-on engagement. They are embedded, not imposed, and they equip the child with capacities for clarity, discernment, and agency.

The future is not separate from the child's present; it is held gently in their hands each day within the prepared environment. As guides, our task is to continue preparing ourselves so that we may truly prepare the child.

To support this work, the following overview traces how AI-relevant competencies tend to unfold across developmental stages. This is not a checklist or sequence to be enforced, but a living map intended to support observation, reflection, and thoughtful response. Children acquire these capacities asynchronously, according to readiness, interest, and experience.

There is no one-size-fits-all approach. Each school and staff community is encouraged to adapt this framework in ways that align with their learners, values, and context.

Chapter 21

Mapping Montessori Materials for AI Competency Development

AI Competency Development by Age Group

The following overview presents the full *Mapping Montessori Materials for AI Competency Development* structure, highlighting where key capacities are typically introduced and how they deepen over time. These competencies do not develop in a linear or uniform manner. Rather, they are refined gradually as children engage with materials, environments, and challenges that meet them where they are:

AI Competency Development	0 to 3	3 to 6	6 to 12	12 to 18	Currently Integrated in Montessori
Social Intelligence & Empathy					Yes
Order & Sequence Awareness					Yes
Spatial Awareness					Yes
Pattern Recognition					Yes
Cause & Effect					Yes
Hand-eye Coordination					Yes
Tactile & Sensory Data Processing					Yes
Simple Problem-solving Steps					Yes
Order & Symmetry					Yes
Motor Skills & Coordination					Yes
Algorithmic Thinking Foundations					Yes
Observational Skills					Yes
Data Sorting & Classification					Yes
Attention to Detail					Yes
Digital Citizenship as Practical Life					Partially
Precision & Control					Yes
Sequential Logic					Yes
Basic Computational Thinking					Yes
Conceptual Understanding of Patterns					Yes
Basic Data Analysis					Yes
Problem-solving with Complexity					Partially
Algorithm Creation					Partially
Ethical Considerations					Partially
Mathematical Foundations in AI					Partially
Indirect Preparation for Natural Language Processing					Partially
Advanced Computational Thinking					Partially
Data Structuring & Analysis					Partially
Data Visualization					Partially
Algorithm Optimization					Partially
Machine Learning Foundations					Partially
Project-based AI Applications					Partially
AI Ethics & Self-Reflection					Partially

KEY: Academic Levels

Infant / Toddler (0-3)
Early Childhood (3-6)
Elementary (6-12)
Adolescence (12-18)

Montessori Materials Mapping For AI Competency Development Chart

Here is how each AI competency might manifest in a Montessori classroom:

0-3 Years

- **Social Intelligence & Empathy** – Grace & courtesy lessons, observing facial expressions, mirroring emotions.
- **Order & Sequence Awareness** – Arranging materials on a tray, following a simple routine, setting a table.
- **Spatial Awareness** – Nesting cups, stacking blocks, crawling through tunnels, object permanence work.
- **Pattern Recognition** – Matching colors, sorting by size, simple sound games, responsive caregiving.
- **Cause & Effect** – Rattles, tactile mobiles, rolling a ball down a ramp, objective permanence boxes.
- **Hand-eye Coordination** – Transferring snacks with tongs, threading beads (infilare materials), placing pegs into a board (imbucare materials).
- **Tactile & Sensory Data Processing** – Touch boards, fabric matching, mystery bag exploration, sensory bins.
- **Simple Problem-solving Steps** – Experiments in nature, texture matching, treasure baskets, mystery bag.

3-6 Years

- **Order & Symmetry** – Arranging pink tower, knobbed cylinders, geometric cabinet, setting up workspaces.
- **Motor Skills & Coordination** – Pouring grains, using tweezers, buttoning and zipping frames, clay modeling.
- **Algorithmic Thinking Foundations** – Step-by-step tasks, moving objects in sequence, bead stairwork.
- **Observational Skills** – Silence game, nature walks, object matching games, storytelling with visual cues.
- **Data Sorting & Classification** – Sorting shells, leaf or animal cards, geometric shapes, classification charts.
- **Attention to Detail** – Sandpaper letters, mystery bag, binomial cube, identifying small changes in materials.
- **Digital Citizenship as Practical Life** – Grace & courtesy around devices, discussions on responsible choices, storytelling with picture sequences.

6-12 Years

- **Precision & Control** – Metal insets, cursive writing practice, bead frame math, scientific experiments.
- **Sequential Logic** – Sentence analysis, long division work, timeline activities, coding unplugged games.
- **Basic Computational Thinking** – Using decimal system materials, pattern extension exercises, storytelling with logic-based choices.
- **Conceptual Understanding of Patterns** – Fibonacci spiral exploration, tessellations, geometry in art.
- **Basic Data Analysis** – Graphing favorite activities, recording weather patterns, tracking plant growth.
- **Problem-solving with Complexity** – Open-ended math problems, practical life troubleshooting, bridge-building challenges.
- **Algorithm Creation** – Writing simple instructions for a friend to follow, map-making, coding board games.
- **Ethical Considerations** – Group discussions on fairness, responsibility in digital spaces, moral dilemma stories.

12-18 Years

- **Mathematical Foundations in AI** – Probability exercises, matrices, statistics, real-world data applications.
- **Indirect Preparation for Natural Language Processing** – Sentence diagramming, language deconstruction, poetry analysis.
- **Advanced Computational Thinking** – Hands-on coding projects, logic puzzles, designing functional robots.
- **Data Structuring & Analysis** – Spreadsheet work, cataloging library books, analyzing environmental impact data.
- **Data Visualization** – Creating infographics, mapping historical events, using charts in research presentations.
- **Algorithm Optimization** – Refining a process for efficiency, debugging code, improving an experiment's methodology.
- **Machine Learning Foundations** – Training simple AI models (e.g., teaching an AI chatbot), game strategy analysis.
- **Project-based AI Applications** – Creating AI-related research projects, building chatbots, AI in creative arts.
- **AI Ethics & Self-Reflection** – Debates on AI's role in society, ethical case studies, personal reflection journals on technology use.

Now that we have explored how AI competencies can be understood developmentally, let us look more closely at how these abstract skills are already present, often beautifully and quietly, in the Montessori environment. The following section offers tangible examples of how specific competencies might unfold across age groups, paired with corresponding Montessori materials and practices. This is not a prescriptive list, but a living map, one that invites continued observation, reflection, and refinement as educators bring their own insight and creativity to this work.

Chapter 22

0–3 Years Infant / Toddler

In the Infant and Toddler community, the focus is on sensory exploration, motor coordination, and foundational social-emotional development. Competencies such as spatial awareness, cause and effect, and hand-eye coordination are **already** cultivated through real-world, hands-on activities like sorting, stacking, and practical life tasks. These foundational experiences build neural connections essential for early cognitive and physical growth. The Montessori approach for this age group remains screen-free, prioritizing engagement with the physical environment to nurture independence, curiosity, and sensory integration.

While digital citizenship is not explicitly taught at this stage, its foundations are embedded in practical life, sensorial work, and peace education; it focuses on establishing physical experience, respect for self, others, and the environment. Lessons in grace and courtesy, turn-taking, and respectful handling of materials form the early building blocks for respectful behavior in digital spaces as children grow. By grounding these behaviors in real-life interactions, children develop empathy and responsibility, which later translate into mindful digital engagement.

Chapter 23

0–3 Social Intelligence & Empathy

Curriculum Area: Practical Life / Social-Emotional Development
Subject / Strand: Screen-Free Digital Citizenship Foundations
Lesson Title: Social Intelligence & Empathy (Infant/Toddler 0–3)
Age Range: 0–3 years

Prerequisites / Previous Work:
No prerequisites. This work is designed for the ongoing development of emotional literacy and social engagement from infancy onward.

Materials:

- Puppets and soft toys for emotional role-play
- Family photo cards to support recognition and relational bonding
- Grace and Courtesy lessons modeled in context
- A simple mirror for observing self-expression
- Baby dolls with accessories for nurturing routines such as feeding and bathing

Presentation:
Prepare a calm, orderly environment, free of screen distractions, where materials are attractively displayed within the child's reach. Invite the child into the space with warmth and clarity. Introduce

one material at a time, using clear, simple language. For example: 'This is a puppet. It helps us show how we feel.' Model the material slowly and deliberately, breaking movements into distinct steps. Use the puppet, for instance, to express basic emotions such as happy, sad, or excited. Then offer the material to the child and step back, allowing for free exploration. Observe carefully. Offer prompts only if needed (e.g., "What feeling is the puppet showing us?"). Let the child engage at their own pace and with their own emotional rhythm.

Language:
Use emotion-based vocabulary appropriate to the child's level: happy, sad, excited, tired, gentle, kind. Introduce terms naturally during interactions.

Points of Interest:
Children often become captivated by mirrors and faces, their own or those in photos. Their natural drive to imitate gestures, care for others, and respond to feelings is enhanced through these materials.

Direct Aims:
To support the development of empathy and social intelligence through observation, imitation, and interaction. To offer concrete ways for the child to recognize and express emotions.

Indirect Aims:
To build self-awareness, nurture fine motor control in caregiving, and strengthen the foundation for future communication and collaboration.

Control of Error:
Misalignments between emotional cues and responses guide the adult in offering repeated modeling. Child's disengagement may indicate the need to simplify or vary the presentation.

Extensions / Variations:
Use dolls or puppets to act out simple social interactions such as sharing or comforting. Introduce routines such as feeding or wrapping baby dolls. Read picture books focused on feelings and social problem-solving, pausing to reflect: "How do you think she feels?"

Follow-up Work / Assessment:
Observe the child's spontaneous engagement with the materials. Note recognition and response to facial expressions or emotions. Track social interactions, including turn-taking, offering comfort, or using empathetic gestures like patting a peer.

Notes for the Guide:
The foundation of digital citizenship begins with human connection. These early expressions of empathy and care support emotional literacy long before screens are introduced.

Chapter 24

0–3 Order & Sequence Awareness

Curriculum Area: Practical Life / Cognitive Development
Subject / Strand: Screen-Free Digital Citizenship Foundations
Lesson Title: Order & Sequence Awareness (Infant/Toddler 0–3)
Age Range: 0–3 years

Prerequisites / Previous Work:
No prior experience required. This work supports natural developmental tendencies for order and exploration during the absorbent mind phase.

Materials:

- Stacking blocks for spatial awareness and size differentiation
- Nesting cups to explore volume and containment
- Knobbed cylinders for visual discrimination and graded ordering
- Sorting trays with objects of different shapes, sizes, or colors
- Real-life objects such as measuring spoons or matching lids and containers

Presentation:
Prepare a calm, tidy environment, ensuring all materials are attractively arranged within the child's view and reach. Invite the

child with warmth and guide them to the work area. Present the material using clear, simple language (e.g., "We are going to stack these blocks from biggest to smallest"). Demonstrate slowly and deliberately, breaking the activity into distinct steps. As you stack or nest each item, name its size or position (e.g., "This is the largest cup; next is the smaller one"). Offer the materials to the child and step back to observe. Let the child engage at their own pace, exploring the order intuitively. Offer gentle prompts only if needed, such as "What comes next?" or "Is this bigger or smaller?" to support their process without directing it.

Language:
Use spatial and comparative terms such as big, small, biggest, next, last, top, bottom, same, different. Introduce language naturally during exploration and repetition.

Points of Interest:
Children are often drawn to size differences, repetition, and fitting objects together. These materials provide satisfying outcomes when completed correctly, supporting the child's inner drive for order.

Direct Aims:
To support the development of ordering and sequencing skills through hands-on manipulation and observation
To promote logical thinking, concentration, and problem-solving

Indirect Aims:
To prepare the child for later concepts such as pattern recognition, mathematics, and algorithmic thinking
To build habits of attention, organization, and independence

Control of Error:
Materials are designed to fit only in specific sequences. A mismatch may frustrate or confuse the child, prompting self-correction or repeated attempts.

Extensions / Variations:
Invite the child to collect natural objects such as leaves, stones, or flowers and sort them by size, color, or shape. Offer simple visual patterns (e.g., red-blue-red) using sorting items or blocks and encourage the child to replicate or continue the pattern. Demonstrate real-life sequences such as setting the table, buttoning a shirt, or following steps in a dressing routine.

Follow-up Work / Assessment:
Observe the child's focus, repetition, and ability to follow or create sequences. Look for signs of spontaneous organization and problem-solving, as well as independent correction.

Notes for the Guide:
Early sequencing supports future competencies in logic, mathematics, and even coding. This foundational work strengthens executive function, giving children tools to navigate an increasingly structured, tech-influenced world.

Chapter 25

0–3 Spatial Awareness

Curriculum Area: Sensorial / Cognitive Development
Subject / Strand: Screen-Free Digital Citizenship Foundations
Lesson Title: Spatial Awareness (Infant/Toddler 0–3)
Age Range: 0–3 years

Prerequisites / Previous Work:
No prerequisites required. This work builds naturally on the infant's movement and exploration phase, supporting body-object awareness and sensorimotor coordination.

Materials:

- Shape sorters to develop matching and placement skills
- Simple puzzles for visual discrimination and hand-eye coordination
- Block-building sets to foster understanding of balance and spatial relationships, this progresses to the Pink Tower in the 3-6 environment.
- Baskets of varied objects for tactile and spatial exploration
- Tunnels or low obstacle courses to support gross motor navigation and directionality

Presentation:
Prepare a clean, quiet space with materials neatly arranged at

the child's level. Invite the child with warmth and guide them to the shelf or table. Introduce the chosen material with simple language (e.g., "This is a shape sorter. Let's see how the shapes fit inside."). Demonstrate slowly and deliberately, naming each shape while fitting it into the matching slot. Pause between actions to allow observation. Offer the material to the child and step back, observing their process without interference. Allow independent exploration and creativity, only offering gentle prompts if necessary (e.g., "Does the triangle fit here?"). Respect the child's effort and pace throughout.

Language:
Use shape and positional vocabulary: triangle, circle, square, under, over, beside, between, taller, smaller. Introduce new words gradually during play and repetition.

Points of Interest:
Children are naturally intrigued by fitting things together, stacking, and crawling through spaces. The satisfaction of successful shape-matching or balanced stacking enhances intrinsic motivation.

Direct Aims:
To support the development of spatial awareness and body-in-environment orientation
To encourage problem-solving and visual discrimination through interactive materials

Indirect Aims:
To prepare for future mathematical and geometric reasoning
To enhance coordination, concentration, and motor planning

Control of Error:
Materials are self-correcting, shapes only fit in their corresponding slots, and blocks fall if unbalanced. The child naturally adjusts actions based on outcomes.

Extensions / Variations:
Invite the child to build structures with blocks, then replicate or modify them. Use positional vocabulary during movement games (e.g., “Can you walk around the table?”). Take the child on short exploration walks, noting spatial relationships in the environment (e.g., “The tree is behind the bench.”). Offer varied objects to sort, stack, or explore in different spaces.

Follow-up Work / Assessment:
Observe how the child matches and manipulates shapes or puzzle pieces. Note problem-solving strategies, persistence, and focus. Look for emerging use of spatial vocabulary in daily interactions or imitation.

Notes for the Guide:
Spatial awareness underpins future skills in engineering, robotics, and design. These early embodied experiences allow the child to build mental models of the physical world, laying the groundwork for abstract thinking and creative construction in the years to come.

Chapter 26

0-3 Pattern Recognition

Curriculum Area: Sensorial / Early Mathematics
Subject / Strand: Screen-Free Digital Citizenship Foundations
Lesson Title: Pattern Recognition (Infant/Toddler 0–3)
Age Range: 0–3 years

Prerequisites / Previous Work:
No prior work required. This lesson naturally follows sensorial exploration, object permanence, and matching experiences commonly observed in early development.

Materials:

- Colored blocks or beads for creating visual patterns
- Repeating design objects such as striped balls, dotted fabrics, or patterned trays
- Sequencing cards with simple two- or three-step visual patterns
- Nature items like shells, leaves, stones, or petals for outdoor sorting and sequencing
- Musical instruments such as bells, rhythm sticks, or shakers for auditory patterns

Presentation:
Prepare the environment by placing materials neatly on a low shelf

or surface accessible to the child. Ensure the space is calm and free from noise or visual distractions. Invite the child with a warm greeting and introduce the activity clearly (e.g., "We are going to make a pattern with these beads"). Begin by demonstrating a simple repeating pattern using two or three objects—such as "red, blue, red"—and name each color aloud as you place it. Move slowly and deliberately to encourage focused observation. Offer the materials to the child and invite them to explore or replicate the pattern at their own pace. Observe closely. Offer gentle prompts only when needed (e.g., "What comes next?") without correcting or interfering with the child's creative process. Celebrate discovery and variation.

Language:
Use pattern vocabulary naturally: red, blue, again, next, same, repeat, sound, quiet, jump, clap, rhythm. Incorporate names of shapes, colors, and sounds when appropriate.

Points of Interest:
Children are drawn to rhythm, repetition, and symmetry. Patterns appeal visually and kinesthetically, and successful repetition brings satisfaction and confidence.

Direct Aims:
To develop the child's ability to recognize, follow, and create simple patterns using visual, tactile, and auditory cues
To lay the foundation for early mathematical thinking and cognitive sequencing

Indirect Aims:
To support concentration, memory, and prediction skills
To prepare the child for logical reasoning, early coding, and musical rhythm

Control of Error:
Patterns that break or lose rhythm stand out clearly, encouraging the child to self-correct. When a piece doesn't match the sequence,

the child may naturally adjust without external correction.

Extensions / Variations:
Collect natural items and create outdoor patterns (e.g., stone-leaf-stone). Explore sound sequences by clapping, tapping, or using instruments in a rhythmic order (e.g., "clap-clap-tap"). Introduce movement patterns such as "jump, step, jump" and encourage the child to follow, lead, or invent their own. Offer opportunities to create with fabric textures or food items, introducing patterns through meal prep or snack arranging.

Follow-up Work / Assessment:
Observe the child's ability to recognize, replicate, or extend patterns. Take note of verbal or physical cues when responding to sequences. Watch for moments of experimentation, joy, or persistence in arranging materials in logical or creative ways.

Notes for the Guide:
Pattern recognition supports higher-order thinking across disciplines. These early interactions foster skills required for future fields such as mathematics, music, programming, and visual design, where understanding and building on patterns is essential.

Chapter 27
0–3 Cause & Effect

Curriculum Area: Sensorial / Cognitive Development
Subject / Strand: Screen-Free Digital Citizenship Foundations
Lesson Title: Cause & Effect (Infant/Toddler 0–3)
Age Range: 0–3 years

Prerequisites / Previous Work:
No prerequisites required. This lesson builds upon the child's natural desire to explore, interact with objects, and observe outcomes through movement and sound.

Materials:

- Rattles for auditory feedback through shaking
- Objects with levers, switches, or buttons to trigger visual or sound-based responses
- Pop-up toys that activate through pressing or pulling
- Light-up or musical objects that react to touch
- Water play items such as cups, water wheels, and funnels to demonstrate fluid motion and effect

Presentation:
Prepare a safe, calm workspace and arrange the materials neatly within the child's reach. Invite the child with warmth and guide them to the activity. Use simple, expressive language to explain

what will happen (e.g., “Listen to what happens when I shake this rattle!”). Demonstrate one item slowly, clearly showing the connection between movement and reaction. For example, shake the rattle and pause to hear the sound, or press a button and observe a light. Emphasize both the action and the result. Then offer the item to the child and allow space for independent exploration. Observe closely, stepping in only to reinforce the cause-effect relationship as it occurs (e.g., “You pushed it, and the light turned on!”). Ask simple, reflective questions to spark curiosity without leading.

Language:
Use everyday cause-and-effect phrases such as: when, because, push, shake, press, it lights up, it moves, it makes sound. Encourage naming the action and result.

Points of Interest:
Children are naturally fascinated by sound, light, and motion. The predictable feedback from these materials provides satisfaction, surprise, and motivation for repetition.

Direct Aims:
To support the child’s ability to understand and predict outcomes from actions
To foster early reasoning, problem-solving, and exploration

Indirect Aims:
To strengthen observation skills, focus, and logical thinking
To prepare the child for future scientific inquiry and mechanical exploration

Control of Error:
Built into the materials, if no result occurs, the child may change their action or attempt a new method. This invites experimentation and refinement of understanding.

Extensions / Variations:
Offer water play items such as wheels, cups, or pitchers and observe how water moves when poured or redirected. Use rolling balls or toy cars to explore motion, speed, and slopes. Lift-the-flap or sound books provide immediate visual or auditory outcomes from a simple action, reinforcing the concept of reaction.

Follow-up Work / Assessment:
Observe how the child interacts with cause-and-effect materials. Look for repetition, excitement, and anticipation. Note problem-solving behaviors such as trying new approaches to recreate a result. Document changes in attention span, engagement, and emerging understanding of how actions lead to specific reactions.

Notes for the Guide:
Understanding cause and effect builds the foundation for future critical thinking and scientific reasoning. These early experiences with active learning shape how the child perceives relationships, systems, and interactions in the wider world, skills vital in technology-rich environments.

Chapter 28

0–3 Hand-Eye Coordination

Curriculum Area: Practical Life / Motor Development
Subject / Strand: Screen-Free Digital Citizenship Foundations
Lesson Title: Hand-Eye Coordination (Infant/Toddler 0–3)
Age Range: 0–3 years

Prerequisites / Previous Work:
No prerequisites required. This lesson builds upon the child's early reaching and grasping reflexes and introduces more intentional, coordinated actions.

Materials:

- Threading beads to develop fine motor precision and visual tracking
- Stacking rings for aligning shapes and encouraging controlled placement
- Large-piece puzzles to support matching, fitting, and grip strength
- Scooping and pouring sets for coordinated, fluid movements
- Wooden or fabric-covered blocks to practice stacking and balance

Presentation:

Create a calm, inviting environment by placing the selected materials neatly within the child's reach. Eliminate distractions to maintain the child's focus. Invite the child with warm curiosity and without expectation, using simple, encouraging language (e.g., "We are going to thread the bead onto the string"). Begin the demonstration by slowly and deliberately completing the action step by step. Emphasize how the hands and eyes work together, narrating gently (e.g., "Watch how I hold the string and slide the bead on"). If the child struggles, gently guide their hands with minimal intervention, offering just enough support to build confidence. Avoid completing the activity for them. Allow the child to repeat the motion independently. Observe with interest and patience. Repeat demonstrations only if necessary, always modeling calm, focused movements.

Language:
Use motor-focused terms such as thread, scoop, pour, hold, slide, fit, balance. Reinforce these words naturally during the activity without over-instruction.

Points of Interest:
Children are often fascinated by how things fit, balance, or fall. The challenge of mastering hand control sparks concentration and repetition, which supports deeper learning.

Direct Aims:
To support the development of fine motor control through hand-eye coordination
To build concentration, precision, and independence in movement

Indirect Aims:
To prepare the child for future writing, dressing, feeding, and creative work
To strengthen problem-solving, perseverance, and spatial awareness

Control of Error:
Built-in feedback from the materials—if a bead misses the string or a block topples, the child can see and adjust. These natural cues promote self-correction and resilience.

Extensions / Variations:
Introduce pouring activities with small pitchers or spooning dry materials like rice or beans into a bowl. Offer snack prep tasks such as spreading or slicing soft fruits with child-safe tools. Encourage movement-based relays where the child transfers items from one point to another using their hands or a spoon. Provide crayons or chalk for drawing simple shapes, reinforcing visual tracking and intentional motion.

Follow-up Work / Assessment:
Observe how the child grasps, aligns, and manipulates materials. Look for increasing precision, fewer spills, and smoother motions. Track persistence and repeated attempts as signs of developing mastery.

Notes for the Guide:
Hand-eye coordination is the bridge between intention and action. Strengthening it through purposeful, joyful work sets the foundation for independence in everyday life and success in academic and creative pursuits that require dexterity and focus.

Chapter 29

0-3 Tactile & Sensory Data Processing

Curriculum Area: Sensorial / Cognitive Development
Subject / Strand: Screen-Free Digital Citizenship Foundations
Lesson Title: Tactile & Sensory Data Processing (Infant/Toddler 0–3)
Age Range: 0–3 years

Prerequisites / Previous Work:
No prerequisites required. This lesson builds naturally from early touch-based exploration and supports integration of the senses through hands-on experiences.

Materials:

- Texture boards featuring smooth, rough, bumpy, or fuzzy surfaces
- Sandpaper letters for early tactile exposure to symbols
- Sensory bins filled with rice, beans, water beads, or natural materials
- Fabric swatches of various materials such as silk, cotton, burlap, or velvet
- Clay or playdough to mold, press, and shape with hands

Presentation:
Prepare a calm, distraction-free workspace with sensory materials arranged neatly and within reach. Invite the child to the activity

with warm curiosity. Use simple, descriptive language to introduce the materials (e.g., "Let us feel these textures with our hands"). Begin by gently touching one of the materials yourself and narrating your experience (e.g., "This feels rough. My fingers can feel all the bumps"). Hand the material to the child and invite exploration. Encourage them to touch, press, or move their fingers over the surfaces. Use vocabulary to describe sensations and prompt comparison when appropriate (e.g., "Is this one softer than that one?"). Allow the child to explore freely, offering new textures gradually to sustain engagement. Follow their interests and avoid interruption unless guidance is needed.

Language:
Use sensory-focused terms such as soft, rough, smooth, bumpy, sticky, cold, warm, squishy, hard. Introduce new vocabulary naturally during exploration.

Points of Interest:
Children are drawn to variety in texture, temperature, and form. Materials that offer contrast invite deeper exploration and repeat interactions, fostering attention and curiosity.

Direct Aims:
To support the development of sensory processing through tactile engagement
To encourage vocabulary development through descriptive sensory language

Indirect Aims:
To build fine motor coordination, self-regulation, and emotional awareness
To prepare the child for future academic skills, including writing and scientific observation

Control of Error:
The child's own senses guide the experience. Mismatches or unexpected sensations prompt reflection, curiosity, or adjusted interaction.

Extensions / Variations:
Offer warm and cool objects such as stones, spoons, or cloths for temperature comparison. Use a mystery bag game where the child identifies familiar objects by touch. Take a nature walk and explore textures such as tree bark, rocks, grass, and leaves. Ask reflective questions like, "What does this feel like?" or "Have you felt something like this before?"

Follow-up Work / Assessment:
Observe the child's willingness to explore various textures. Note emotional responses such as excitement, hesitation, or joy. Track vocabulary use and independent comparisons. Repetition and engagement indicate increasing sensory awareness and confidence.

Notes for the Guide:
Sensory integration lays the foundation for learning, emotional regulation, and physical coordination. These tactile experiences support the child's capacity to process a complex world, preparing them for tasks requiring calm focus, adaptability, and precise control in both physical and digital environments.

Chapter 30

0-3 Simple Problem-Solving Steps

Curriculum Area: Practical Life / Cognitive Development
Subject / Strand: Screen-Free Digital Citizenship Foundations
Lesson Title: Simple Problem-Solving Steps (Infant/Toddler 0–3)
Age Range: 0–3 years

Prerequisites / Previous Work:
No prior work required. This lesson builds naturally on exploration and trial-and-error experiences already present in the child's environment.

Materials:

- Shape sorters for visual matching and spatial reasoning
- Stackable materials such as nesting cups or rings to support sequencing and coordination
- Puzzles with a few large pieces to build logical thinking and patience
- Practical life items like small containers with lids or twist tops
- Wooden or foam blocks for creative problem-solving and experimentation with balance

Presentation:
Set up a calm, distraction-free environment with materials arranged

neatly and accessibly. Invite the child warmly and approach the work area with curiosity and positivity. Present the task in clear, simple language (e.g., "Let us figure out where this shape fits"). Slowly demonstrate how to approach the problem, using verbal cues to describe your process (e.g., "This is round. Let's see if it fits in the round space"). Keep movements slow and purposeful. Hand the material to the child and allow for independent exploration. Observe closely. If the child encounters difficulty, offer minimal guidance, such as "What do you think happens if we turn it this way?" Always focus on effort rather than outcome. Acknowledge attempts and persistence with encouragement and warmth.

Language:
Use words and phrases like try, turn, fit, doesn't work, try again, you figured it out, what else can you do, let's solve this together. Offer language that names both the challenge and the thinking process.

Points of Interest:
Children are naturally drawn to puzzles, matching games, and fitting tasks. The process of discovering how things work provides motivation and builds inner satisfaction.

Direct Aims:
To support the development of independent thinking and persistence
To offer experiences that promote logical reasoning through hands-on problem-solving

Indirect Aims:
To strengthen confidence, resilience, and adaptability
To build the foundation for future scientific inquiry and creative innovation

Control of Error:
Materials offer immediate feedback, if the piece doesn't fit or the tower falls, the child naturally receives information that prompts a

new attempt.

Extensions / Variations:
Introduce multi-step tasks, such as stacking blocks in a specific sequence to complete a structure. Use nature-based materials to build uneven towers or sort items by type. Offer real-world tasks like opening a jar, tying a soft string, or putting on shoes, all of which involve planning and effort.

Follow-up Work / Assessment:
Observe how the child responds to challenge and novelty. Note signs of problem-solving, such as trying again after a failed attempt or using a new method. Look for moments of self-correction, engagement, and joy after success.

Notes for the Guide:
Simple problem-solving builds the mental habits children will rely on in future collaborative, creative, and tech-integrated environments. These early moments of persistence, exploration, and discovery set the stage for confidence in tackling increasingly complex tasks.

Chapter 31

3-6 Years Early Childhood

In the Early Childhood environment, competencies such as logical reasoning, order and sequence, and attention to detail are **already** developed through practical life, sensorial, and math activities. Children practice early problem-solving and data classification using materials like the Pink Tower, Brown Stairs, and bead chains. Role-playing and Grace and Courtesy lessons also support the development of social intelligence and empathy. This level continues to emphasize screen-free learning, as tactile materials and real-world interactions foster both fine motor skills and executive function. Extensions often connect learning to nature-based observations, further reinforcing Montessori's emphasis on grounding learning in reality.

Digital citizenship continues as both practical life and peace studies at this level by focusing on fairness, sharing, and respect for community spaces and tools. Through collaborative tasks, children learn to navigate social dynamics, practice patience, and resolve conflicts peacefully. These lessons instill values of kindness and responsibility that later inform their understanding of responsible digital behavior.

Chapter 32
3-6 Order & Symmetry

Curriculum Area: Sensorial / Early Mathematics
Subject / Strand: Screen-Free Digital Citizenship Foundations
Lesson Title: Order & Symmetry (Early Childhood 3–6)
Age Range: 3–6 years

Prerequisites / Previous Work:
Familiarity with basic sensorial materials such as the Pink Tower, Broad Stair, and Color Tablets. This work deepens earlier sensorial experiences and builds readiness for abstract reasoning.

Materials:

- Pattern blocks for symmetrical design and geometric arrangements
- Geometric solids to explore form and spatial relationships
- Pink Tower to develop gradation and visual order
- Brown Stair for sequencing by size and weight
- Constructive Triangles to support symmetry, shape formation, and visual balance

Presentation:
Prepare a calm and aesthetically pleasing workspace. Display the materials in an orderly and inviting manner on low shelves. Invite the child with enthusiasm and introduce the activity in simple,

clear language (e.g., "We are going to build something beautiful from biggest to smallest"). Begin by demonstrating the Pink Tower, stacking cubes from largest to smallest. Draw attention to the visual and physical balance. Transition to using pattern blocks to create a symmetrical design, describing each step aloud (e.g., "Now I place a yellow hexagon in the middle, and a red trapezoid on each side"). Emphasize the idea of balance and repetition. Invite the child to try, observing their choices and use of space. Ask reflective questions such as "What do you notice about your pattern?" or "Does each side look the same?" Offer vocabulary that supports their discoveries.

Language:
Use terms such as balance, symmetry, equal, biggest, smallest, same, opposite, matching, center, repeat, and order. Introduce vocabulary in context, emphasizing clarity and beauty.

Points of Interest:
Children are drawn to beauty and balance. Symmetry offers visual satisfaction and encourages repetition and refinement. The sense of "rightness" in symmetrical designs promotes concentration and pride in work.

Direct Aims:
To develop awareness of spatial relationships, balance, and order
To support precision, pattern recognition, and early mathematical thinking

Indirect Aims:
To foster aesthetic appreciation and attention to detail
To prepare for future work in geometry, architecture, and design

Control of Error:
Built into the materials, mismatched sizes or unbalanced arrangements are visually evident. This invites the child to adjust and refine independently.

Extensions / Variations:
Encourage collecting natural objects such as leaves, rocks, or flower petals to explore natural symmetry. Offer templates for symmetrical drawings like mandalas or folding activities. During outdoor walks, point out examples of symmetry in nature and architecture. Invite the child to recreate symmetrical arrangements from memory or observation.

Follow-up Work / Assessment:
Observe the child's engagement and ability to create balanced, intentional designs. Note their use of symmetry in new contexts and interest in aligning or organizing materials. Look for growing independence and joy in creating precise, beautiful work.

Notes for the Guide:
Order and symmetry help children see the underlying structure of the world. These skills are foundational to creative expression, mathematical reasoning, and spatial intelligence. The child who sees balance in a flower or structure is preparing to design, engineer, and appreciate beauty in all forms.

Chapter 33

3–6 Motor Skills & Coordination

Curriculum Area: Practical Life / Physical Development
Subject / Strand: Screen-Free Digital Citizenship Foundations
Lesson Title: Motor Skills & Coordination (Early Childhood 3–6)
Age Range: 3–6 years

Prerequisites / Previous Work:
Basic familiarity with self-care routines and simple hand movements. This lesson builds on earlier sensorial exploration and hand-eye coordination work from the toddler years.

Materials:

- Practical life tools such as tweezers, tongs, and small pitchers to build grip strength and precision
- Threading beads to refine finger dexterity and bilateral coordination
- Child-safe scissors for cutting practice and muscle control
- Scooping and pouring sets for developing smooth, coordinated movement
- Cloth folding exercises to support structure, repetition, and purposeful movement

Presentation:
Create a clean, organized workspace. Lay out each material neatly

on a tray or mat, ensuring items are in the child's reach. Invite the child warmly and explain the task using clear, simple language (e.g., "Today, we will practice moving objects using these tongs"). Maintain an encouraging tone throughout. Demonstrate slowly and deliberately, breaking the task into clear, manageable steps. Show how to hold the tool correctly, and model the movement with exaggerated calm and control (e.g., transferring one object from a bowl to another using tongs). Emphasize precision and fluidity in each action. Offer the tool to the child and observe. Provide gentle verbal prompts if needed (e.g., "Can you try moving the bead again slowly?"), but allow space for independent discovery. Encourage repetition, supporting the development of coordination and muscle memory. Celebrate all progress with curiosity and encouragement.

Language:
Use practical terms such as grip, scoop, pour, fold, cut, hold, slow, careful, strong, and gentle. Integrate language naturally into observation and conversation during the activity.

Points of Interest:
Children often enjoy mastering tools and tasks that mirror real-life adult activities. The sense of purpose and responsibility embedded in these movements increases motivation and focus.

Direct Aims:
To develop fine and gross motor coordination through purposeful movement
To strengthen concentration, independence, and physical confidence

Indirect Aims:
To prepare for future tasks such as writing, typing, cutting, and tool use
To support sequencing, attention to detail, and body awareness

Control of Error:
Materials provide direct feedback. Spills, slips, or uneven folding signal the need for adjustment and allow the child to self-correct through observation and experience.

Extensions / Variations:
Increase the challenge with smaller tools or more delicate objects. Introduce multi-step tasks such as setting the table or preparing a snack. Use creative tasks like drawing patterns, cutting shapes, or forming letters from playdough to blend coordination with expression.

Follow-up Work / Assessment:
Observe the child's fluidity and accuracy in movement. Note increasing independence and confidence. Watch for perseverance, especially when a task is repeated without prompting.

Notes for the Guide:
Strong motor coordination builds the foundation for independence in self-care, creative expression, and academic engagement. Supporting these skills through meaningful, hands-on activities reinforces the child's confidence in using their body to explore, create, and contribute.

Chapter 34

3-6 Algorithmic Thinking Foundations

Curriculum Area: Mathematics / Practical Life
Subject / Strand: Screen-Free Digital Citizenship Foundations
Lesson Title: Algorithmic Thinking Foundations (Early Childhood 3–6)
Age Range: 3–6 years

Prerequisites / Previous Work:
Familiarity with sequencing in daily routines or step-based activities such as handwashing or basic counting. Builds upon early pattern recognition and order.

Materials:

- Montessori math materials including Number Rods, Golden Beads, and Cards & Counters
- Sequencing cards depicting everyday actions (e.g., getting dressed, preparing food)
- Practical life materials for multi-step activities (e.g., table setting, flower arranging)
- Building blocks with visual or verbal step-by-step guidance

Presentation:
Prepare a clean, quiet space with materials laid out neatly and in order. Invite the child with enthusiasm and introduce the concept simply (e.g., “We are going to put these pictures in the right

order to show how we get ready in the morning"). Present the sequencing cards and guide the child through placing them in a logical order. Narrate the steps aloud (e.g., "First we wake up, then we get dressed"). As you arrange, ask the child to notice familiar patterns or repeated actions. Encourage them to reflect (e.g., "Do we always wash our hands before we eat?"). Provide a new set of sequencing cards or manipulatives and invite the child to complete the task independently. Observe quietly, offering encouragement and feedback without correcting unless asked. Celebrate the child's reasoning and discoveries, emphasizing their thinking process.

Language:
Use process-based words such as first, next, then, last, after, repeat, plan, steps, and order. Use this vocabulary to describe both familiar routines and abstract sequences.

Points of Interest:
Children naturally enjoy identifying routines and solving step-by-step problems. These materials and tasks offer a satisfying sense of order and predictability.

Direct Aims:
To support the child's ability to think sequentially and logically
To build a foundation for planning, problem-solving, and pattern recognition

Indirect Aims:
To prepare the child for future computational thinking, including coding and procedural logic
To strengthen memory, self-direction, and confidence in approaching multi-step tasks

Control of Error:
Out-of-order steps become apparent through visual cues or storytelling gaps, inviting natural self-correction and reflection.

Extensions / Variations:
Create movement sequences like "clap, clap, step" and have the child repeat or extend them. Use instruments to build sound sequences. Offer puzzles that must be assembled in a specific order. Use story sequencing cards to place events in order, then have the child retell the story aloud. Introduce more complex daily routines and challenge the child to plan them independently.

Follow-up Work / Assessment:
Observe the child's sequencing process and ability to reflect on the logic of their choices. Note persistence, adaptability, and willingness to revise their arrangement. Track emerging interest in routines, instructions, and planning.

Notes for the Guide:
Algorithmic thinking lays the foundation for organizing thoughts and solving problems across disciplines. These early steps into planning and sequencing empower the child to approach challenges methodically, developing lifelong skills in logic, creativity, and adaptability.

Chapter 35

3-6 Observational Skills

Curriculum Area: Sensorial / Science
Subject / Strand: Screen-Free Digital Citizenship Foundations
Lesson Title: Observational Skills (Early Childhood 3–6)
Age Range: 3–6 years

Prerequisites / Previous Work:
Experience with sensorial exploration and nature-based activities. Builds upon the child's natural curiosity and interest in the environment.

Materials:

- Magnifying glasses to enhance close visual observation
- Nature observation tools such as binoculars, tweezers, or collection jars
- Art supplies including drawing paper, crayons, and colored pencils
- Montessori Botany Cabinet for comparing and classifying leaf shapes
- Natural collections such as rocks, shells, feathers, or pinecones

Presentation:
Prepare a quiet and inviting space indoors or outdoors with

materials displayed neatly. Invite the child to the activity with enthusiasm and explain simply (e.g., "We're going to look closely at something special and see what we can find"). Present an object like a leaf, shell, or flower, and introduce the magnifying glass. Model how to use it slowly and carefully, pointing out visible details (e.g., "Look at the lines on this leaf—it's like a little road map"). Invite the child to take their time observing. Ask open-ended questions to spark curiosity (e.g., "What colors do you see?" or "How does it feel?"). Provide art materials and invite the child to draw what they observed or describe it aloud. Guide them to compare their drawing with the object, highlighting new details or features.

Language:
Use descriptive and sensory language such as shape, line, edge, smooth, rough, pointy, round, color, and pattern. Encourage children to use their own words and expand their vocabulary through conversation.

Points of Interest:
Children are often drawn to the small, overlooked elements of nature and everyday objects. Tools like magnifying glasses and art materials invite deeper exploration and wonder.

Direct Aims:
To develop focused attention and detail-oriented observation
To support descriptive language and early scientific thinking

Indirect Aims:
To build curiosity, memory, and comparison skills
To prepare for data collection, critical analysis, and artistic expression

Control of Error:
Overlooked details can be rediscovered through repeated observation or reflection, often prompted by drawing or conversation

Extensions / Variations:
Create a personal nature journal to record findings over time. Use the Botany Cabinet to compare real leaves with their matching geometric shapes. Invite the child to describe similarities and differences as they explore leaf shapes. Go on observational walks and document changes in the environment across seasons. Encourage comparisons between new findings and past entries.

Follow-up Work / Assessment:
Observe the child's use of vocabulary and visual detail in their artwork or descriptions. Watch for emerging patterns in curiosity, such as returning to the same spot or object, or asking deeper questions.

Notes for the Guide:
Observational skills form the basis of scientific inquiry and creative awareness. Helping children slow down and truly see the world around them cultivates wonder, care, and thoughtfulness, essential qualities for meaningful learning in both scientific and humanistic disciplines.

Chapter 36

3–6 Data Sorting & Classification

Curriculum Area: Sensorial / Early Mathematics
Subject / Strand: Screen-Free Digital Citizenship Foundations
Lesson Title: Data Sorting & Classification (Early Childhood 3–6)
Age Range: 3–6 years

Prerequisites / Previous Work:
Experience with sensorial materials and pattern work. Builds on the child's natural tendency to categorize and compare through exploration and conversation.

Materials:

- Sorting trays for organizing objects by attribute
- Attribute blocks for sorting by color, shape, and size
- Everyday items such as buttons, toy animals, or leaves with varying features
- Montessori Color Tablets to explore shades and hues
- Nature collections including rocks, shells, pinecones, or dried flowers

Presentation:
Set up a clean, distraction-free workspace with materials arranged in trays or baskets. Invite the child with curiosity and introduce the task in clear, simple language (e.g., "Let us sort these objects

by color"). Choose one attribute and model the process step by step." These are all red, and these are blue." Emphasize the common characteristic of each group and clearly name the attributes. Then invite the child to try sorting independently. Observe their sorting process and offer encouragement (e.g., "You noticed all the squares!"). Once the child is confident sorting by one attribute, introduce a second (e.g., "Now, let us sort by shape or size"). Engage in reflective dialogue about the groups they've made (e.g., "Which group has more? How are these alike?"). Offer vocabulary and insight without correcting unless clarification is requested.

Language:
Use sorting-related terms like group, same, different, size, color, shape, big, small, more, fewer, set, match, and category. Introduce classification terms through natural conversation during exploration.

Points of Interest:
Children enjoy organizing, matching, and discovering patterns. The clarity and logic of sorting satisfies their developmental need for order while offering playful engagement.

Direct Aims:
To build foundational skills in sorting, grouping, and classifying information
To support logical reasoning and early data analysis

Indirect Aims:
To prepare the child for future understanding of categorization in mathematics, science, and technology
To build vocabulary, attention to detail, and flexible thinking

Control of Error:
The attributes themselves guide the child, mismatched items stand out visually or tactilely. The act of sorting naturally reveals inconsistencies.

Extensions / Variations:
Introduce Venn diagrams using hoops or circles on paper to explore overlapping categories (e.g., items that are both red and round). Collect natural objects during a walk and sort them by observable traits. Create sorting games with time challenges or verbal prompts (e.g., "Find everything that is small and smooth!"). Invite the child to create their own sorting rules.

Follow-up Work / Assessment:
Observe how the child chooses to group items and how many attributes they consider. Look for creativity in category-making and flexibility in re-sorting. Note their language, focus, and whether they initiate similar activities independently.

Notes for the Guide:
Sorting and classification develop the mental structure for data literacy. These skills build the foundation for understanding how systems organize information—an essential competency in the digital age. Montessori's hands-on approach brings abstract data concepts into the child's tangible, playful world.

Chapter 37

3-6 Attention to Detail

Curriculum Area: Sensorial / Practical Life
Subject / Strand: Screen-Free Digital Citizenship Foundations
Lesson Title: Attention to Detail (Early Childhood 3–6)
Age Range: 3–6 years

Prerequisites / Previous Work:
Experience with matching, sorting, or sensorial materials. Builds on the child's developing visual discrimination and fine motor control.

Materials:

- Knobbed Cylinders for precise placement and size matching
- Matching cards with images, patterns, or silhouettes
- Color Tablets for sorting subtle shade variations
- Sorting trays with nuanced differences in object features
- Practical life tools for tasks like water pouring, polishing, or spooning small objects

Presentation:
Prepare a peaceful, well-organized workspace with the chosen materials laid out neatly and within the child's reach. Invite the child to the work area and introduce the task with intention (e.g., "Let us see how carefully we can match these cylinders to their correct spaces"). Demonstrate slowly, using calm and

deliberate movements. As you work, draw attention to subtle details (e.g., "This one fits exactly because it's just the right size—too big and it won't go in"). Show care in handling the materials, modeling mindfulness and precision. Hand the materials to the child and allow them to explore independently. Observe quietly. If they encounter difficulty, offer gentle prompts that support self-correction (e.g., "Does that look like it fits the same way?"). Acknowledge effort and improvement.

Language:
Use descriptive terms such as same, different, smaller, tighter, lighter, smoother, exact, and careful. Embed vocabulary naturally in context as the child observes and interacts.

Points of Interest:
Children are drawn to puzzles and tasks that challenge their accuracy. The satisfaction of a perfect match or correctly completed sequence provides intrinsic motivation.

Direct Aims:
To enhance focus, visual discrimination, and fine motor precision
To support self-correction and thoughtful engagement with materials

Indirect Aims:
To prepare for writing, art, and technical tasks that require care and observation
To build internal discipline, pride in work, and awareness of process

Control of Error:
Materials are self-correcting, incorrect placements stand out visually or functionally. The child is encouraged to refine through trial and error.

Extensions / Variations:
Offer matching games using textured or patterned objects. Provide

shadow-matching cards or memory-style games. Use creative projects such as mosaics, coloring within detailed outlines, or constructing symmetrical designs. Explore nature together, looking closely at the fine features of leaves, insects, or stones. Invite the child to draw what they observe.

Follow-up Work / Assessment:
Observe the child's level of concentration and care. Note persistence when making corrections or improvements. Track improvements in precision and control over time. Look for moments of pride or reflection on their process.

Notes for the Guide:
Attention to detail nurtures both cognitive and creative capacities. By supporting the child's ability to notice, refine, and appreciate subtle differences, we build habits that serve them across disciplines, from careful writing to design, from scientific observation to expressive art.

Chapter 38

3-6 Digital Citizenship as Practical Life

Curriculum Area: Practical Life / Grace and Courtesy
Subject / Strand: Screen-Free Digital Citizenship Foundations
Lesson Title: Digital Citizenship as Practical Life (Early Childhood 3–6)
Age Range: 3–6 years

Prerequisites / Previous Work:
Basic experience with Grace and Courtesy lessons, Practical Life routines, and simple sorting or role-play activities. Builds upon natural social interactions and peer modeling.

Materials:

- Wooden cards or tokens symbolizing digital tools (e.g., sharing, permission, safety)
- Visual aids showing behaviors like helping, asking, and waiting turns
- Simple circuitry kits or unplugged coding tools for collaborative tasks
- Icon cards representing kind vs. unsafe behavior
- Role-play props such as aprons, tools, or toy phones to simulate social scenarios

Presentation:
Create a calm, welcoming space with symbolic materials neatly arranged. Invite the child and begin by connecting digital concepts to familiar actions (e.g., "When we ask to use crayons, we're practicing respect—just like we would if we wanted to share something online"). Use analog tools to explore concepts like permission, sharing, and kindness. Demonstrate sorting symbolic cards into categories such as "kind," "safe," and "needs work," explaining each choice with clear examples. Introduce simple wiring kits or coding puzzles to model how connections require cooperation, just like respectful digital interactions. Offer hands-on sorting and invite the child to try classifying actions themselves. Guide role-play scenarios where children say "please," "thank you," or ask permission, reinforcing empathy and respect in both digital and face-to-face settings.

Language:
Use real-world and metaphorical terms such as kind, share, respect, safe, ask, connect, helpful, harmful, responsible. Make abstract ideas accessible through familiar actions and shared language.

Points of Interest:
Children enjoy storytelling, role-play, and games that reflect real-life behavior. Symbolic materials and hands-on tasks help them connect ethical choices to practical experience.

Direct Aims:
To introduce foundational ideas of digital citizenship through hands-on, respectful, real-world experiences
To foster early understanding of collaboration, safety, and kindness in shared spaces

Indirect Aims:
To prepare the child for future online interaction with strong social and emotional grounding
To support the development of empathy, problem-solving, and

ethical awareness

Control of Error:
Group dialogue and role-play provide opportunities for self-correction. Sorting tasks and feedback reinforce awareness of choices and consequences.

Extensions / Variations:
Tell stories about characters practicing kindness and responsibility in digital-like scenarios. Set up a human "circuit" by connecting hands to demonstrate how respect keeps communication flowing. Challenge children to create their own symbols for helpful or harmful digital tools and explain them. Use sorting games with more complex scenarios to encourage critical thinking.

Follow-up Work / Assessment:
Observe how the child identifies respectful behavior, responds to prompts, or initiates role-play. Note language use, empathy, and signs of internalizing respectful routines. Look for spontaneous application of lessons in peer interactions.

Notes for the Guide:
Digital citizenship begins with grace, empathy, and thoughtful action. By embedding these values in Practical Life activities, we prepare children to navigate both the real and digital world with care, curiosity, and responsibility.

Chapter 39

6-12 Years Elementary

At the Elementary level, the curriculum broadens to include interdisciplinary learning that **already** fosters competencies such as critical thinking, collaboration, and sequential logic. Students work with advanced Montessori materials like the Checkerboard for multiplication and the Timeline of Life for integrated studies in history and biology. They engage in project-based learning that encourages both individual exploration and group work, nurturing computational thinking and ethical reasoning. Limited or optional screen use may be introduced for research, simple coding exercises, or data visualization to enhance inquiry. However, digital tools are integrated thoughtfully to support, rather than replace, hands-on exploration.

Digital citizenship at this stage is framed as both practical life and peace studies, emphasizing responsible research, data privacy, and respectful communication during collaborative work. Students are guided to evaluate sources critically, respect intellectual property, and practice safe online interactions, reinforcing a sense of digital responsibility. This aligns with peace studies by fostering cooperation, mutual respect, and an understanding of the ethical dimensions of their actions within digital and real-world communities.

Chapter 40

6–12 Precision & Control

Curriculum Area: Practical Life / Geometry / Art
Subject / Strand: Screen Use as Practical Life
Lesson Title: Precision & Control (Elementary 6–12)
Age Range: 6–12 years

Prerequisites / Previous Work:
Experience with basic measuring tools, early sewing or art work, and sensorial foundations in geometry or handwriting. Builds upon earlier lessons in coordination and concentration.

Materials:

- Geometric Stick Material for constructing and measuring precise figures
- Constructive Triangles to explore and replicate geometric relationships
- Sewing kits with child-safe needles, embroidery hoops, and fabric
- Woodworking tools including small hammers, nails, and measuring instruments
- Calligraphy or brush-lettering supplies including ink, brushes, and quality paper

Presentation:
Prepare a quiet, organized workspace with all materials displayed neatly. Invite the child to the work area and introduce the activity with intention (e.g., “Today, we’ll explore how to draw a perfect circle using this compass”). Discuss how developing precision helps in creating clean, accurate, and beautiful results. Demonstrate the task slowly and with care, narrating each step (e.g., “I place the point here, keep it steady, and turn gently to make an even circle”). Emphasize posture, grip, and control. Invite the child to try, and observe their process without interruption. Provide constructive support if needed (e.g., “Try steadying your wrist as you draw”), while celebrating effort and progress. Encourage repetition for refinement and confidence building.

Language:
Use terminology such as steady, measure, align, control, curve, accurate, angle, stroke, refine, and consistent. Offer clear, supportive language that connects the child’s effort with improvement.

Points of Interest:
Children enjoy mastering tools that produce tangible, high-quality results. The satisfaction of a symmetrical drawing or a neatly sewn stitch encourages repetition and self-motivation.

Direct Aims:
To build accuracy, coordination, and attention to fine detail
To develop confidence in using tools and materials with care and intention

Indirect Aims:
To strengthen focus, perseverance, and craftsmanship
To prepare for technical and artistic disciplines that require meticulous execution

Control of Error:

The outcome reflects the child's precision, uneven stitches, wobbly circles, or incorrect angles become visible indicators for self-correction and refinement.

Extensions / Variations:
Invite the child to create detailed geometric designs using multiple shapes and measurements. Introduce calligraphy or brush-lettering exercises with increasing complexity. Explore real-world applications by assembling a model, crafting a precise design for a science project, or creating a visual timeline or chart.

Follow-up Work / Assessment:
Observe the child's willingness to repeat the task and attention to improving technique. Note increasing accuracy and pride in their finished work. Look for a growing sense of ownership and independence in their use of precision tools.

Notes for the Guide:
Precision and control support both academic mastery and creative expression. These skills foster confidence in handling challenges with patience and care, critical qualities in any screen-integrated or hands-on work. By linking this competency to real-life applications, we nurture discipline, artistry, and innovation.

Chapter 41

6–12 Sequential Logic

Curriculum Area: Mathematics / Practical Life / History
Subject / Strand: Screen Use as Practical Life
Lesson Title: Sequential Logic (Elementary 6–12)
Age Range: 6–12 years

Prerequisites / Previous Work:
Experience with early math operations, timeline work, or multi-step practical life tasks. Builds upon basic understanding of order, patterns, and planning.

Materials:

- Montessori math materials: Long Division Boards, Multiplication Bead Frames, and the Stamp Game
- Coding cards with visual instructions for unplugged programming exercises
- Sequential puzzles, mazes, or logic-based board games
- Historical timeline materials and event cards
- Multi-step practical life materials for tasks like baking, assembling, or organizing

Presentation:
Prepare a quiet, organized workspace with all materials arranged in sequence. Invite the child with interest and introduce the goal

(e.g., "Today we'll work through a long division problem by breaking it into steps"). Demonstrate the process clearly, modeling how to think step by step (e.g., "First, we divide. Then, we multiply. Next, we subtract…"). Use consistent, clear language to highlight the logical flow of actions. Invite the child to take over, guiding them with reflective prompts such as "What comes next?" and "How do you know that's the next step?" Encourage pauses for review and adjustment if needed. Let the child explore and complete the full process independently.

Language:
Use sequencing terms such as first, next, then, finally, step-by-step, sequence, process, instructions, and order. Reinforce vocabulary as part of reflective conversation.

Points of Interest:
Children often enjoy uncovering solutions through step-based challenges. The satisfaction of a completed sequence supports confidence and encourages repetition.

Direct Aims:
To strengthen logical thinking and step-by-step problem-solving
To support independent planning and execution of multi-step processes

Indirect Aims:
To develop resilience, patience, and cognitive flexibility
To prepare for advanced concepts in coding, science, and history

Control of Error:
Steps followed out of order often lead to incorrect outcomes, prompting self-correction. Math and coding materials are self-verifying through outcomes and processes.

Extensions / Variations:
Introduce unplugged coding exercises using cards or manipulatives.

Create scavenger hunts that require completing clues in order. Engage in practical projects like baking or building that involve sequential steps. Use historical timelines or science experiments to show step-based progression in real-world contexts.

Follow-up Work / Assessment:
Observe how the child handles sequences—do they plan, pause, revise, and reflect? Note persistence when solving puzzles or completing tasks. Ask the child to explain their process or invent their own step-by-step activity.

Notes for the Guide:
Sequential logic provides a framework for thinking clearly in increasingly complex situations. By supporting children in practicing and applying these steps in practical, academic, and imaginative contexts, we empower them to become confident planners, problem-solvers, and designers of their own learning paths.

Chapter 42

6–12 Basic Computational Thinking

Curriculum Area: Mathematics / Logical Thinking
Subject / Strand: Screen Use as Practical Life
Lesson Title: Basic Computational Thinking (Elementary 6–12)
Age Range: 6–12 years

Prerequisites / Previous Work:
Familiarity with step-based tasks, pattern work, or basic math operations. Builds upon earlier skills in sequencing, categorization, and logic.

Materials:

- Bead Frames for step-by-step math practice and place value exploration
- Hands-on coding kits with physical programming pieces or logic blocks
- Flowchart templates for planning processes and mapping decisions
- Sequencing cards for constructing logical sequences and instructions
- Logic games or puzzles requiring ordered steps and pattern recognition

Presentation:
Prepare a clean, organized workspace with a variety of materials available. Invite the child and introduce the purpose of the work (e.g., "We're going to solve problems step by step, like building a clear set of instructions"). Begin with a relatable example such as organizing a morning routine or preparing a snack. Demonstrate how to use a flowchart to map each decision and step. Use clear, simple language to model decision-making. For example: 'If we are hungry, we move to this step. If not, we go here.' Encourage the child to try with a different real-life task, supporting them in breaking it into parts, identifying patterns, and making decisions. As they grow confident, invite them to try independent activities using coding kits, sequencing cards, or puzzles.

Language:
Introduce and reinforce terms such as step, sequence, pattern, decision, input, output, repeat, and logic. Embed language naturally during demonstrations and conversations.

Points of Interest:
Children are naturally curious about solving problems and predicting outcomes. Computational thinking transforms everyday experiences into creative, structured challenges that children can explore and master with confidence.

Direct Aims:
To introduce step-by-step thinking and logical problem-solving
To build early familiarity with algorithms, sequencing, and system-based approaches

Indirect Aims:
To prepare for future learning in programming, robotics, and systems thinking
To develop perseverance, strategic thinking, and reflective problem-solving

Control of Error:
Logical processes and puzzles reveal success or missteps clearly through the outcome. The child receives direct feedback from the accuracy and efficiency of their steps.

Extensions / Variations:
Engage the child in pattern recognition games and challenge them to create patterns for others to solve. Invite them to write algorithms for daily tasks such as setting the table or planting a seed. Offer logic puzzles or riddles that require methodical steps and reasoning. Encourage collaboration with peers to debug or refine created processes.

Follow-up Work / Assessment:
Observe the child's approach to breaking down challenges. Track how they respond to errors. Do they revise thoughtfully? Note use of language around logic and systems. Look for increased confidence in creating and applying sequential instructions independently.

Notes for the Guide:
Computational thinking is not just for technology, it is a life skill. It empowers children to approach challenges with clarity, confidence, and creativity. With repeated practice, they learn to view complex systems as solvable puzzles, preparing them for both academic success and real-world problem-solving.

Chapter 43

6–12 Conceptual Understanding of Patterns

Curriculum Area: Mathematics / Geometry / Cultural Studies
Subject / Strand: Screen Use as Practical Life
Lesson Title: Conceptual Understanding of Patterns (Elementary 6–12)
Age Range: 6–12 years

Prerequisites / Previous Work:
Familiarity with sequences, basic arithmetic, or early geometry. Builds upon early sensorial and visual discrimination work, as well as foundational number concepts.

Materials:

- Racks & Tubes for observing numerical patterns in division
- Geometric Stick Material for exploring symmetry, tessellations, and polygon relationships
- Binomial and Trinomial Cubes for identifying algebraic patterns and expansions
- Fibonacci materials or natural pattern cards showing spirals, seed arrangements, or shells
- Cultural artifacts or images (e.g., textiles, mosaics, architectural elements) for cross-disciplinary pattern recognition

Presentation:
Prepare a visually rich, orderly space with mathematical and artistic materials ready for hands-on use. Invite the child to explore patterns in the world around them (e.g., “Today we’ll discover how patterns can help us understand math, nature, and even art”). Begin with a familiar material, such as the Racks & Tubes, and demonstrate how repeated steps in division reveal consistent numerical patterns. Ask the child to describe what they see and predict what might come next. Transition to geometric or cultural materials and encourage identification of visual patterns. Use real-world examples, such as petals on flowers or tiles in a building, to spark connections between disciplines.

Language:
Use terms such as repeat, sequence, symmetry, tessellation, spiral, pattern, rhythm, and relationship. Reinforce these words through observation, creation, and discussion.

Points of Interest:
Patterns are deeply satisfying to children—offering predictability, beauty, and the thrill of discovery. From math to design, noticing how elements repeat and relate provides insight and excitement.

Direct Aims:
To recognize, extend, and create mathematical, visual, and natural patterns
To support logical reasoning and artistic interpretation through structured repetition

Indirect Aims:
To build a foundation for data analysis, coding, design, and scientific observation
To develop the child’s appreciation for structure, order, and creative variation

Control of Error:
Patterns either continue logically or break when misapplied. Materials like beads, shapes, or steps in a sequence make inconsistencies easily visible for self-correction.

Extensions / Variations:
Invite children to design patterns using colored pencils or mosaic tiles. Explore cultural examples of patterns in architecture, textiles, and ceramics, and encourage recreation or reinterpretation. Go on nature walks to identify and document naturally occurring patterns in leaves, flowers, or tree bark. Challenge the child to invent a new type of pattern and explain how it works.

Follow-up Work / Assessment:
Observe how the child identifies, describes, and extends patterns. Note curiosity, originality, and how confidently they apply pattern recognition in new tasks. Encourage discussion about why patterns matter and where they appear in their world.

Notes for the Guide:
Understanding patterns is central to how humans interpret data, express creativity, and solve problems. By integrating pattern work across subjects, we give children tools for thinking structurally, acting creatively, and engaging with the world more deeply.

Chapter 44

6–12 Basic Data Analysis

Curriculum Area: Mathematics / Science / Practical Life
Subject / Strand: Early Screen Use as Practical Life
Lesson Title: Basic Data Analysis (Elementary 6–12)
Age Range: 6–12 years

Prerequisites / Previous Work:
Experience with recording observations, using measurement tools, and basic graphing. Builds upon prior experience with number sense, categorization, and pattern recognition.

Materials:

- Graphing tools: rulers, graph paper, pencils, markers
- Charts and tables: pre-made or blank templates for recording data
- Data collection kits: thermometers, timers, measuring tapes, rain gauges
- Decimal Checkerboard for exploring numerical groupings and place value
- Age-appropriate digital tools for creating basic graphs and visualizations

Presentation:
Prepare a clear, organized workspace with tools and materials

neatly arranged. Invite the child to join and explain the purpose of the activity (e.g., "We're going to collect and organize information so we can look for patterns and answer questions"). Present a simple dataset, such as temperature readings over a week, and model how to record and plot it on a chart or graph. Use guiding questions such as, "What do you notice?" and "Can we predict what might happen next?" Invite the child to explore their own interests through data collection, such as measuring plant growth, tracking reading time, or observing weather conditions. Encourage them to organize this information using bar graphs, line charts, or tally tables.

Language:
Use data-related vocabulary such as record, measure, analyze, trend, pattern, chart, compare, graph, and conclusion. Encourage the child to explain their findings using descriptive language.

Points of Interest:
Children enjoy discovering meaningful insights through real-world data. Seeing numbers come alive in a chart or graph makes information more concrete and relevant to their lives.

Direct Aims:
To develop the ability to collect, organize, and interpret numerical and observational data
To build confidence in drawing conclusions and making predictions based on evidence

Indirect Aims:
To prepare the child for future applications in science, economics, and digital technologies
To foster analytical thinking, logic, and critical reflection

Control of Error:
Charts, graphs, and visualizations provide clear feedback, errors in measurement or recording become evident when data doesn't align. Self-correction is built into the process.

Extensions / Variations:
Guide the child to collect data from personal habits or classroom routines. Compare historical or scientific datasets with their own findings. Introduce simple digital graphing tools and let the child create different visual formats for the same dataset. Encourage journal reflections on their discoveries.

Follow-up Work / Assessment:
Observe the child's ability to record and interpret data independently. Look for growing interest in real-world questions and self-directed projects. Note the clarity of their visualizations and ability to explain their conclusions.

Notes for the Guide:
Basic data analysis cultivates habits of inquiry, accuracy, and evidence-based reasoning. These skills empower children to ask meaningful questions and use logic to solve problems, an essential foundation for navigating the modern world, from scientific investigation to responsible screen use.

Chapter 45

6–12 Problem-Solving with Complexity

Curriculum Area: Practical Life / Mathematics / STEM
Subject / Strand: Early Screen Use as Practical Life
Lesson Title: Problem-Solving with Complexity (Elementary 6–12)
Age Range: 6–12 years

Prerequisites / Previous Work:
Experience with multi-step processes, logic puzzles, and collaborative tasks. Builds upon earlier exposure to sequencing, cause and effect, and practical planning.

Materials:

- Problem-solving kits (e.g., STEM challenge sets, building blocks, engineering cards)
- Montessori math materials: Binomial Cube, Trinomial Cube, Algebraic Squares
- Logic puzzles including tangrams, mazes, or strategic board games
- Practical Life tools for activities like sewing, woodworking, and baking
- Geometric Stick Material for structural design, symmetry, and spatial reasoning

Presentation:

Set up an organized, inspiring workspace with a variety of complex, hands-on challenges available. Invite the child and explain the purpose (e.g., "Today we're going to solve a problem that takes multiple steps, and maybe a few tries!"). Choose a material or puzzle and begin by modeling how to observe and break the challenge into parts. Use guiding questions like, "What do we need to figure out first?" or "What's one possible way to begin?" Emphasize the value of patience and perseverance. After your brief demonstration, hand the materials to the child and step back, observing their approach. Encourage reflection as they progress, and offer supportive feedback without providing direct solutions.

Language:
Use strategic vocabulary such as step-by-step, adjust, test, revise, retry, solution, obstacle, and progress. Encourage the child to verbalize their process and solutions using this language.

Points of Interest:
Children are drawn to puzzles and challenges that feel like real-world mysteries. Multi-step problems stimulate curiosity and often result in spontaneous collaboration and inventive thinking.

Direct Aims:
To build the child's ability to approach complex tasks methodically and creatively
To strengthen resilience, persistence, and critical reasoning through real-world challenges

Indirect Aims:
To prepare the child for long-term projects, teamwork, and interdisciplinary thinking
To support the development of flexible problem-solving strategies that apply to both academic and personal challenges

Control of Error:
Built into the structure of each challenge, unsuccessful attempts provide immediate feedback, guiding the child toward refinement and alternate strategies.

Extensions / Variations:
Create group challenges that require collaboration, such as building a bridge, organizing an event, or creating a shared game. Present real-life problems (e.g., planning a community space or classroom layout) and invite the child to design and iterate solutions. Encourage invention through design-based tasks with no fixed outcome.

Follow-up Work / Assessment:
Observe how the child approaches uncertainty and setbacks. Note creativity in developing multiple solutions, willingness to revise, and ability to work independently or with peers. Track the child's use of planning tools and how they verbalize their thought process.

Notes for the Guide:
Complex problem-solving is a foundational skill for innovation, leadership, and personal growth. Montessori environments already nurture persistence and deep thinking, by naming and supporting this competency, we equip children to confidently navigate an evolving world full of challenges and opportunities.

Chapter 46

6–12 Algorithm Creation

Curriculum Area: Practical Life / Mathematics / Computational Thinking
Subject / Strand: Early Screen Use as Practical Life
Lesson Title: Algorithm Creation (Elementary 6–12)
Age Range: 6–12 years

Prerequisites / Previous Work:
Experience with sequencing, procedural thinking, and step-based tasks. Builds on earlier competencies in problem-solving, pattern recognition, and logical reasoning.

Materials:

- Coding blocks or tiles representing loops, actions, and conditional statements
- Flowchart templates and drawing tools for mapping steps visually
- Sequential task cards illustrating everyday procedures
- Practical life tools for real-life algorithms (e.g., setting a table, making a recipe)
- Logic puzzles that require ordering steps or identifying rules

Presentation:
Set up a neat, engaging workspace with materials clearly organized.

Invite the child and introduce the concept by saying, "An algorithm is a list of steps that tells us how to complete a task." Use a relatable example, such as making a sandwich, and demonstrate writing or drawing each step in order. Emphasize the importance of sequence (e.g., "If we add the peanut butter before opening the jar, the task won't work!"). Then, invite the child to choose a familiar routine—like brushing teeth or packing a bag—and guide them in creating their own algorithm. Encourage testing the algorithm as written, identifying what works and what needs adjusting. Support self-correction and celebrate iteration.

Language:
Use terms such as steps, sequence, repeat, loop, condition, refine, and instruction. Encourage the child to use clear and specific language when writing or describing their steps.

Points of Interest:
Children enjoy organizing and breaking down real-life challenges into manageable parts. Algorithm creation invites ownership, creativity, and clarity in thinking.

Direct Aims:
To develop the ability to create and follow step-by-step instructions
To strengthen logical sequencing and problem-solving through practical tasks

Indirect Aims:
To lay the foundation for future coding, robotics, and STEM exploration
To support planning, organization, and clear communication in everyday life

Control of Error:
The outcome of the task provides feedback, if the algorithm is incomplete or out of order, the child sees the result and is invited to adjust. Testing reinforces self-correction.

Extensions / Variations:
Invite the child to code their algorithms using age-appropriate platforms or apps. Present design challenges such as, “Write an algorithm to build a tower with 5 blocks using only one hand.” Encourage creating algorithms for morning routines, art projects, or organizing a shelf. Turn algorithms into team activities where one child writes the steps and another follows them.

Follow-up Work / Assessment:
Observe the child’s ability to organize steps logically, respond to errors, and refine their instructions. Note engagement, creativity, and independence. Look for clear verbal or written articulation of their process.

Notes for the Guide:
Algorithmic thinking is foundational in both digital systems and daily decision-making. By treating screen-related logic as a hands-on practical life skill, children develop confidence and fluency in sequencing, adaptability, and problem-solving, competencies vital for thriving in both analog and digital environments.

Chapter 47

6–12 Ethical Considerations

Curriculum Area: Practical Life / Grace and Courtesy / Social Studies
Subject / Strand: Early Screen Use as Practical Life
Lesson Title: Ethical Considerations (Elementary 6–12)
Age Range: 6–12 years

Prerequisites / Previous Work:
Experience with collaborative work, group discussions, or role-play. Builds upon foundational social-emotional learning and grace and courtesy lessons.

Materials:

- Discussion prompts focused on fairness, responsibility, and community
- Short stories or fables with ethical dilemmas
- Role-play scripts or props to reenact situations requiring moral judgment
- Community meeting tools (e.g., talking stick, discussion guidelines)
- Philosophy inquiry cards for extended ethical dialogue and critical thinking

Presentation:
Create a calm, inclusive discussion area with materials organized and inviting. Welcome the child or group and introduce a scenario (e.g., "What should someone do if they see another child being left out?"). Read a short story or describe a relatable ethical dilemma aloud. Allow time for reflection, then ask open-ended questions such as, "What would you do? Why?" or "Was there more than one right answer?" Guide the conversation toward deeper understanding of empathy, fairness, and responsibility. Support children in expressing their own ideas while listening respectfully to others. Use follow-up questions to explore consequences and motivations. Encourage optional role-play or journaling for children who prefer expressive or quiet reflection.

Language:
Incorporate moral vocabulary such as fairness, empathy, respect, impact, responsibility, harm, help, truth, choice, and consequence. Invite children to define and reflect on these concepts through examples and lived experiences.

Points of Interest:
Children often feel empowered when given space to think deeply about justice, kindness, and personal responsibility. Engaging in ethical reflection builds confidence and a sense of belonging.

Direct Aims:
To strengthen ethical reasoning and awareness of the consequences of choices
To develop perspective-taking, empathy, and responsible decision-making

Indirect Aims:
To prepare for collaborative leadership and moral reasoning in digital, academic, and social contexts
To support the development of principled communication, respect for diversity, and global citizenship

Control of Error:
Ethical discussions are not about right or wrong answers, but about clarity of reasoning and openness to others' ideas. The guide supports respectful self-correction by modeling thoughtful inquiry and open dialogue.

Extensions / Variations:
Explore stories of historical figures facing moral decisions. Host student-led ethical debates on classroom issues. Encourage children to develop their own "What would you do?" scenarios and lead a group discussion. Use creative activities such as writing or drawing to explore emotions and consequences related to specific dilemmas.

Follow-up Work / Assessment:
Observe how children engage in dialogue and reflect on ethical choices. Note their ability to consider multiple perspectives and how they respond to disagreements. Look for increasing awareness of the impact of actions on others.

Notes for the Guide:
Ethical considerations are essential in preparing children for responsible digital citizenship, leadership, and innovation. Montessori environments naturally support moral development through community and reflection, this work invites deeper awareness of how values shape behavior and relationships in all spheres of life.

Chapter 48

12-18 Years Adolescence

In the Secondary program, students engage in advanced, self-directed projects that emphasize algorithmic thinking, data structuring, and reflective ethical decision-making. They apply their knowledge through project-based AI applications, coding, and interdisciplinary research, deepening their understanding of real-world challenges. Competencies such as machine learning foundations, data visualization, and ethical reflection are explored through case studies, simulations, creating shelf-ready materials to teach others, and community-based projects. Discerning screen use and exploration using the scientific method are an essential components at this level, allowing students to analyze datasets, build models, and collaborate on digital platforms. This structured integration of technology supports meaningful, responsible engagement with AI tools while reinforcing Montessori's principles of independence and self-directed inquiry.

Digital citizenship is **already** framed as both practical life and peace studies in the Secondary environment by focusing on ethical decision-making, digital collaboration, and content creation. Students reflect on their digital footprint, practice respectful communication in online communities, and explore the impact of AI systems on society. These discussions connect to peace studies by fostering empathy, fairness, and the responsibility to contribute positively to digital and physical communities. This comprehensive approach prepares students for thoughtful leadership in a technology-driven world.

Chapter 49

12–18 Mathematical Foundations in AI

Curriculum Area: Mathematics / Computational Thinking
Subject / Strand: Discerning Screen Use for Research and Creativity
Lesson Title: Mathematical Foundations in AI (Adolescence 12–18)
Age Range: 12–18 years

Prerequisites / Previous Work:
Experience with algebraic expressions, probability, graphing, and basic data analysis. Builds upon earlier exploration of number relationships, functions, and mathematical reasoning applied to real-world contexts.

Materials:

- Graphing tools: graph paper, scientific calculators, digital graphing platforms
- Probability kits: dice, spinners, and playing cards for hands-on experiments
- Advanced Montessori materials: Algebraic Cubes, Binomial and Trinomial Theorem sets, matrix tiles
- Data analysis templates: charts, spreadsheets, and statistical modeling worksheets
- Project-based kits for collecting, organizing, and analyzing real-world data (e.g., environmental or behavioral data)

Presentation:
Set up a quiet, organized environment with physical and digital math tools accessible. Invite the student and introduce the lesson: "Today we will explore together how math helps us understand how AI works, especially through probability, algebra, and data modeling." Begin with a real-world question, such as, 'How can we use math to predict sleep patterns based on last week's data?" Guide the student in choosing one method of modeling, graphing, calculating probabilities, or using an algebraic equation. Work alongside them to structure a data collection method, organize results, and analyze trends. Highlight where math intersects with AI applications, such as pattern recognition or decision trees.

Language:
Introduce and reinforce terms like variable, function, algorithm, probability, trend, model, prediction, optimization, and pattern recognition. Encourage students to use this vocabulary during modeling activities and in their written reflections.

Points of Interest:
Students often become highly engaged when abstract math is linked to technologies they encounter in daily life, such as phone apps or AI agents. Showing how data becomes meaningful through math empowers them to use these skills in innovative ways.

Direct Aims:
To deepen understanding of how algebra, probability, and statistics are used in real-world and AI applications
To strengthen skills in modeling, interpreting data, and communicating findings clearly

Indirect Aims:
To prepare students for AI literacy, systems thinking, and future STEM pathways
To build confidence in mathematical reasoning and evidence-based decision-making

Control of Error:
Mathematical processes and outcomes offer built-in feedback. Inaccurate predictions or incomplete models encourage students to revisit calculations and improve their models based on logical review.

Extensions / Variations:
Guide students in building simple AI models using platforms like Scratch or Teachable Machine. Introduce calculus concepts such as slope and derivatives in the context of optimization. Connect to interdisciplinary work by integrating data analysis into environmental studies, robotics, or digital design.

Follow-up Work / Assessment:
Observe the student's ability to apply mathematical tools to real-life and simulated AI problems. Note clarity of explanation, accuracy in calculations, and willingness to refine their approach. Assess their reflection on the role of math in data-driven technologies.

Notes for the Guide:
Understanding the math behind AI prepares adolescents to move beyond passive users of technology to conscious creators and evaluators. By grounding this learning in project-based inquiry and practical modeling, we equip students to engage critically and ethically in a data-rich future.

Chapter 50

12–18 Natural Language Processing (NLP)

Curriculum Area: Computational Thinking / Language Arts / Ethics
Subject / Strand: Discerning Screen Use for Research and Creativity
Lesson Title: Natural Language Processing (Adolescence 12–18)
Age Range: 12–18 years

Prerequisites / Previous Work:
Experience with grammar and language structures, basic coding or data analysis, and discussions on AI tools. Builds upon earlier competencies in pattern recognition, digital literacy, and ethical reflection.

Materials:

- Text analysis tools for exploring word frequency, sentiment, and structure
- Translation tools (AI-based and human-generated) for comparative accuracy studies
- Python-based coding platforms (e.g., Jupyter Notebook or Google Colab) for NLP basics
- Annotated datasets for pre-processed text analysis and training sample models
- Discussion prompts addressing AI bias, privacy, and multilingualism in NLP

Presentation:
Prepare a digital workspace with text analysis apps, coding tools, and printed language samples. Invite the student and introduce the purpose of the session: "Today we'll explore how computers understand human language, and how we can help them do it better." Begin with a simple NLP application, such as word frequency analysis or text sentiment detection. Demonstrate how to load a sample text and visualize the results. Ask the student to experiment with different texts, poetry, news, or student writing, and observe language patterns. Transition to building a simple script that counts words or tags sentiment. Prompt the student to reflect on how NLP is used in real life and what challenges arise from teaching language to machines.

Language:
Introduce terms such as tokenization, sentiment, syntax, frequency, bias, classification, translation, dataset, and language model. Encourage students to use this vocabulary when describing their process and conclusions.

Points of Interest:
Students are often surprised and intrigued by how AI interprets language, especially when they notice bias, errors, or unexpected associations. The connection between language and coding empowers both creativity and critical analysis.

Direct Aims:
To understand how AI systems interpret and process language
To build foundational skills in text analysis, pattern recognition, and model development

Indirect Aims:
To foster ethical thinking about representation, fairness, and communication in technology
To connect computational tools with human-centered learning and creative expression

Control of Error:
Inaccurate scripts or analysis results reveal logical missteps or data limitations, prompting revision. Biases in AI output serve as discussion points for further exploration and refinement.

Extensions / Variations:
Invite the student to code a sentiment analysis tool and apply it to reviews or current events. Analyze historical texts for shifting tone or vocabulary over time. Explore AI-generated storytelling and discuss applications in journalism, education, or entertainment.

Follow-up Work / Assessment:
Observe the student's process when building, testing, and revising NLP tools. Note creativity in text selection and accuracy in interpretation. Engage in reflective conversation on bias, data ethics, and real-world impacts of language models.

Notes for the Guide:
NLP is a gateway to human-centered AI literacy. It bridges coding, communication, and critical thinking. By grounding this work in real texts and ethical reflection, adolescents gain insight into both the power and limitations of language-based AI tools—equipping them to lead responsibly in a digital world.

Chapter 51

12–18 Advanced Computational Thinking

Curriculum Area: Computational Thinking / STEM Integration
Subject / Strand: Discerning Screen Use for Research and Creativity
Lesson Title: Advanced Computational Thinking (Adolescence 12–18)
Age Range: 12–18 years

Prerequisites / Previous Work:
Experience with basic coding, logical sequencing, robotics, or flowchart creation. Builds upon foundational skills in algorithm development, debugging, and structured problem-solving.

Materials:

- Coding platforms for designing and testing algorithms (e.g., Scratch, Python, Arduino)
- Logic puzzles, mazes, and strategy-based games to encourage problem decomposition
- Robotics kits for building and programming systems (e.g., LEGO SPIKE Prime, VEX)
- Simulation software for modeling real-world processes (e.g., weather, ecosystems, traffic systems)
- Flowchart and diagram tools: whiteboards, sticky notes, or digital apps for mapping logic visually

Presentation:
Prepare a work area with coding tools, robotics materials, and logic challenges. Invite the student and introduce the concept: "Today, we'll work through a real-world problem by thinking like computer scientists, breaking it down into smaller parts and designing a solution step by step." Present a scenario such as managing a traffic light system or creating a mini app. Model how to identify the problem, define inputs and outputs, and map the steps using pseudocode or a flowchart. Encourage the student to design a solution through their chosen platform. Guide them through testing and iterating based on results and feedback. Reflect together on what worked and what they would change.

Language:
Use vocabulary such as input, output, iteration, condition, loop, flow, pseudocode, algorithm, system, simulation, and feedback. Support students in using this language to describe their process and reasoning.

Points of Interest:
Many students enjoy the challenge of solving open-ended problems where there's no single right answer. The joy of debugging, designing, and refining fosters resilience, creativity, and satisfaction.

Direct Aims:
To develop structured problem-solving strategies and algorithmic thinking
To design, test, and refine logical solutions in complex systems

Indirect Aims:
To support confidence in tackling real-world problems with digital tools
To foster flexible thinking, innovation, and peer collaboration in project-based learning

Control of Error:
Digital and physical tools provide immediate feedback when an algorithm fails or a program doesn't run as expected. Students are encouraged to debug, revise, and try again.

Extensions / Variations:
Invite students to design a simple game or app with branching logic. Explore simulations of ecological systems or environmental impacts. Facilitate group robotics challenges requiring collaborative algorithm development and problem-solving under constraints.

Follow-up Work / Assessment:
Observe the student's approach to breaking down problems and designing solutions. Note persistence during challenges, ability to articulate logic, and openness to feedback. Invite the student to present their process and reflect on successes and next steps.

Notes for the Guide:
Advanced computational thinking empowers adolescents to move from consumers to creators of technology. When students learn to deconstruct problems and design thoughtful, testable solutions, they are not only preparing for careers in tech, they are building the mindset of resilient, innovative changemakers for the future.

Chapter 52

12–18 Data Structuring & Analysis

Curriculum Area: Data Literacy / STEM / Computational Thinking
Subject / Strand: Discerning Screen Use for Research and Creativity
Lesson Title: Data Structuring & Analysis (Adolescence 12–18)
Age Range: 12–18 years

Prerequisites / Previous Work:
Experience with spreadsheets, graphing, or basic data collection. Builds on foundational skills in observation, categorization, and number relationships.

Materials:

- Spreadsheet software for organizing, calculating, and visualizing data
- Database tools for storing, sorting, and querying structured data
- Graphing and data visualization platforms for creating charts and infographics
- Real-world datasets for hands-on exploration (e.g., environmental, health, civic, or school-based data)
- Scientific experiment kits for collecting original data and testing hypotheses

Presentation:

Create a digital workspace with spreadsheet software, relevant datasets, and graphing tools accessible. Invite the student and introduce the purpose: "Today we'll learn how to take raw data, organize it, and find patterns that help us make decisions or tell a story." Present a disorganized dataset (e.g., survey responses or weather data) and model how to clean it, adding column headers, sorting entries, and calculating totals or averages. Guide the student to create visualizations of the data, such as bar graphs or pie charts. Facilitate discussion around what the data might reveal and how it connects to real-world issues.

Language:
Use terms such as dataset, variable, correlation, average, trend, outlier, category, filter, graph, and analysis. Encourage students to use this vocabulary when presenting findings or asking questions.

Points of Interest:
Students enjoy uncovering hidden patterns and making sense of messy information. Structuring and analyzing data fosters confidence and purpose, especially when applied to real-world topics.

Direct Aims:
To develop data structuring skills using digital tools
To strengthen analysis and interpretation through meaningful data exploration

Indirect Aims:
To prepare students for academic research, informed citizenship, and careers in data-centric fields
To cultivate critical thinking and decision-making based on evidence

Control of Error:
Inaccurate formulas, empty fields, or visual misrepresentations provide natural points for self-correction. Graphs or summaries

that don't match data prompt students to revisit their structure or calculations.

Extensions / Variations:
Guide students in analyzing their own data from experiments or observations. Encourage them to build databases for tracking real-world issues (e.g., resource use or attendance patterns). Support data storytelling projects where students use visuals and narrative to communicate their findings to peers or the community.

Follow-up Work / Assessment:
Observe the student's process as they structure and clean raw data. Note their ability to explain trends and apply findings. Invite them to reflect on the relevance of their conclusions and how they might use data to inform real-world decisions or design AI tools.

Notes for the Guide:
Data structuring and analysis are essential for navigating a world shaped by information. By engaging students in the practical and creative dimensions of working with data, we prepare them to lead responsibly and thoughtfully in a data-driven future.

Chapter 53

12–18 Data Visualization

Curriculum Area: Data Literacy / Design / Research Communication
Subject / Strand: Discerning Screen Use for Research and Creativity
Lesson Title: Data Visualization (Adolescence 12–18)
Age Range: 12–18 years

Prerequisites / Previous Work:
Experience with graphing, basic data analysis, or working with spreadsheets. Builds upon earlier skills in pattern recognition, data structuring, and visual design.

Materials:

- Graphing tools: graph paper, rulers, and digital graphing software
- Infographic software for layout and data display (e.g., Canva, Piktochart)
- Programming libraries for data visualization (e.g., Python's Matplotlib or Plotly)
- Sample datasets for hands-on practice (e.g., climate data, school surveys, sports stats)
- Art supplies for creating hand-drawn visual data representations

Presentation:

Prepare a digital and analog workspace with datasets, software, and tools available. Invite the student and introduce the concept: "Today we'll learn how to turn raw data into compelling visuals that help others understand a message quickly and clearly." Begin by sharing examples of various visualizations—bar charts, line graphs, scatterplots, and infographics. Discuss when and why each type is used. Demonstrate how to choose and build a visual using a sample dataset. Emphasize clarity, color use, labeling, and layout. Invite the student to explore a dataset of personal interest and create multiple visualizations, comparing their effectiveness.

Language:
Introduce and reinforce terms such as dataset, variable, trend, correlation, axis, scale, infographic, dashboard, and clarity. Encourage students to describe both technical elements and aesthetic choices in their visual designs.

Points of Interest:
Visualizing data transforms abstract numbers into concrete, powerful stories. Students often discover that small design changes—like color or spacing—can dramatically affect how data is interpreted.

Direct Aims:
To build skills in selecting and applying effective visual formats for data
To develop clarity and impact in communicating insights through visual media

Indirect Aims:
To support the student's confidence in public speaking, research presentation, and critical analysis
To foster creativity and audience awareness in science, journalism, and digital communication

Control of Error:

Visual errors such as mislabeled axes, misleading scales, or cluttered design prompt natural reflection. Peer review helps refine design choices for clarity and meaning.

Extensions / Variations:
Invite the student to create an infographic summarizing a research project. Explore interactive tools like Tableau or Flourish to create live dashboards. Analyze data journalism examples and reverse-engineer the visualization choices. Design a data-driven social media post to raise awareness on a topic of interest.

Follow-up Work / Assessment:
Observe how the student selects appropriate visualization types and tools. Note creativity, precision, and ability to explain why certain designs were chosen. Encourage reflection on how visuals influenced understanding and emotional impact.

Notes for the Guide:
Data visualization is a bridge between information and insight. Whether in AI systems, science communication, or public advocacy, it helps learners see and show the story behind the numbers. This work empowers adolescents to become not just informed consumers of data, but ethical, effective storytellers of it.

Chapter 54

12–18 Algorithm Optimization

Curriculum Area: Computational Thinking / Computer Science
Subject / Strand: Discerning Screen Use for Research and Creativity
Lesson Title: Algorithm Optimization (Adolescence 12–18)
Age Range: 12–18 years

Prerequisites / Previous Work:
Experience with basic algorithm design, coding fundamentals, and logical sequencing. Builds upon skills in debugging, data structuring, and performance testing.

Materials:

- Flowchart templates for visualizing algorithm logic and identifying inefficiencies
- Coding platforms (e.g., Python, Java, Scratch) for building and refining algorithms
- Optimization challenge prompts that require reducing complexity and improving performance
- Performance timers or built-in benchmarking tools for tracking changes in speed or efficiency
- Debugging tools and code analysis resources for identifying and resolving errors

Presentation:

Prepare a digital workspace with coding access, flowchart tools, and a sample algorithm ready to test. Invite the student and introduce the concept: "Today we will improve how an algorithm performs, not just that it works, but that it works better." Present a basic algorithm (e.g., a sorting or search algorithm) and walk through its steps using a flowchart. Run the algorithm and discuss any areas that could be streamlined. Guide the student in reworking the algorithm, removing redundancies, reordering steps, or changing the logic. Run both versions and compare the performance, reflecting on speed, clarity, and resource use.

Language:
Use terminology such as optimization, bottleneck, efficiency, iteration, redundancy, runtime, benchmark, complexity, and debug. Encourage students to describe their process using technical and reflective language.

Points of Interest:
Students often find satisfaction in discovering they can make something *better*, not just functional. Optimization work fosters perseverance and critical analysis, especially when speed and accuracy matter.

Direct Aims:
To develop the ability to identify inefficiencies in processes and design solutions
To foster a deeper understanding of how algorithms function and can be refined

Indirect Aims:
To encourage creative thinking, systems analysis, and performance-based revision
To prepare students for future applications in AI, robotics, software design, and logistics

Control of Error:

Comparison of outputs and run times between original and optimized versions allows for self-correction. Testing and feedback loops support refinement through logic and measurable outcomes.

Extensions / Variations:
Offer collaborative optimization challenges such as finding the most efficient route, reducing app load time, or simplifying a game mechanic. Encourage students to apply algorithmic thinking to everyday routines, such as optimizing their daily schedule or improving a classroom process. Explore well-known algorithms used in AI and reflect on why efficiency is critical at scale.

Follow-up Work / Assessment:
Observe the student's ability to revise, test, and explain their changes. Note how they respond to failed optimizations and whether they iterate creatively. Invite students to explain their thought process, reflect on what did not work, and propose revisions.

Notes for the Guide:
Optimization mirrors the deeper Montessori principle of purposeful refinement, taking something functional and transforming it into something elegant, efficient, and meaningful. This competency not only builds mastery in logic and efficiency but prepares students to enter fields where performance and innovation are inseparable, from autonomous vehicles to ethical AI development.

Chapter 55

12–18 Machine Learning Foundations

Curriculum Area: Artificial Intelligence / Data Science / Ethics
Subject / Strand: Discerning Screen Use for Research and Creativity
Lesson Title: Machine Learning Foundations (Adolescence 12–18)
Age Range: 12–18 years

Prerequisites / Previous Work:
Familiarity with coding basics, data collection, and pattern recognition. Builds upon earlier competencies in algorithm design, data structuring, and ethical inquiry in technology.

Materials:

- Datasets: preloaded or student-created (images, text, numbers, survey results)
- Machine learning platforms: interactive tools like Teachable Machine, Machine Learning for Kids, or Scratch ML extensions
- Coding kits: tools for creating basic classification and prediction tasks
- Visualization tools: whiteboards, slide decks, or digital diagrams to map the learning process
- Experimentation sandboxes: virtual environments for testing models and adjusting parameters

Presentation:
Set up a workspace with access to machine learning platforms and visual tools. Invite the student and introduce the purpose: "Today we will learn how to teach a machine to recognize patterns, just like we do, by giving it examples and helping it improve." Begin with real-world examples of machine learning in action (e.g., facial recognition, voice assistants). Walk students through each phase: selecting features, labeling data, training the model, and testing it with new examples. Use visual mapping to show the relationship between data input, model adjustment, and prediction. Guide the student in building their own model, testing it, and refining it for better performance.

Language:
Introduce key terms such as machine learning, features, labels, training data, testing data, accuracy, bias, overfitting and generalization, if you want to seed more advanced inquiry. Encourage students to use this vocabulary in reflection and explanation.

Points of Interest:
Students often enjoy seeing a machine "learn" from their inputs and respond accordingly. This gives them a felt sense of agency in shaping intelligent systems, mirroring how humans learn through feedback and refinement.

Direct Aims:
To build foundational understanding of how machine learning systems are created and trained
To develop hands-on skills in creating models, interpreting data, and improving system accuracy

Indirect Aims:
To foster ethical awareness around data bias, fairness, and the societal impacts of AI
To prepare students for deeper engagement in AI development and

data science

Control of Error:
Model performance and prediction results provide built-in feedback. Misclassifications, low accuracy, or poor generalization reveal gaps in data balance or feature design.

Extensions / Variations:
Invite students to compare supervised vs. unsupervised models using visual datasets. Encourage creative applications like building a sorting game or a drawing classifier. Analyze case studies of AI failures and successes, discussing causes and consequences. Explore real-world datasets for applications like environmental change or sports statistics.

Follow-up Work / Assessment:
Observe the student's ability to build, test, and refine a model. Assess their understanding of data's role in learning and accuracy. Invite reflective discussion on how bias can enter models and how it can be reduced through better data and design choices.

Notes for the Guide:
Machine learning introduces students to the core of AI systems. By demystifying the training process and illuminating the role of human decision-making in data design, this work cultivates both technical fluency and ethical consciousness, key for empowering future creators and changemakers in a data-driven world.

Chapter 56

12–18 Project-Based AI Applications

Curriculum Area: Artificial Intelligence / Interdisciplinary Project-Based Learning
Subject / Strand: Discerning Screen Use for Research and Creativity
Lesson Title: Project-Based AI Applications (Adolescence 12–18)
Age Range: 12–18 years

Prerequisites / Previous Work:
Familiarity with coding, machine learning basics, and collaborative work. Builds upon previous competencies in data collection, model training, and problem-solving through technology.

Materials:

- Coding platforms (e.g., Python, Scratch, Teachable Machine) for designing AI models
- Datasets (preloaded or student-collected) for real-world testing and refinement
- Project templates to guide workflow: problem statement, data gathering, modeling, evaluation
- Collaboration tools (e.g., Trello, shared docs) for task management and communication
- Presentation supplies: slide decks, infographic tools, and report templates for final project communication

Presentation:

Set up a digital workspace and shared planning environment. Invite the student and introduce the session: "Today, you'll lead a full AI project from idea to solution—working as a team, solving a problem, and reflecting on your process." Begin by identifying a real-world problem that could benefit from AI intervention (e.g., detecting litter through image recognition, analyzing school energy usage, or helping with community resource mapping). Facilitate brainstorming and project scoping. Guide students to define goals, select tools, and plan their workflow with charts or diagrams. They implement, test, and refine their model while assigning team roles and practicing collaborative decision-making. The project concludes with a team presentation focused on both results and learning.

Language:

Introduce and reinforce terms like problem definition, prototype, dataset, model training, debugging, evaluation, collaboration, iteration, innovation, and solution design. Support the use of professional vocabulary in both technical and teamwork contexts.

Points of Interest:

Students are energized by solving problems that matter to them. The project format helps them see how AI can support real-life needs, and they often take pride in being "solution builders."

Direct Aims:

To foster real-world AI project development and teamwork

To strengthen creative problem-solving and project planning through applied learning

Indirect Aims:

To build leadership, communication, and adaptability in technical environments

To develop confidence in managing long-term, interdisciplinary challenges

Control of Error:
The iterative design cycle (plan, build, test, reflect) naturally surfaces issues for self-correction. Collaborative feedback and data-based evaluation foster continuous improvement.

Extensions / Variations:
Students create interdisciplinary projects connecting AI with other subjects (e.g., visualizing climate data in science, creating AI-generated poetry in literature). They may contribute to open-source AI communities or design tools that address a specific need in their school or town.

Follow-up Work / Assessment:
Observe how students define the problem, delegate roles, and engage in troubleshooting. Assess their ability to reflect on outcomes and articulate both technical and collaborative learning. Invite student teams to present findings in a digital format and respond to peer questions.

Notes for the Guide:
AI becomes meaningful when students apply it to their lives and communities. Project-based learning cultivates ethical designers and innovative thinkers who understand both the capabilities and responsibilities of working with intelligent systems. This is where their research becomes real, and their creativity becomes impact.

Chapter 57

12-18 AI Ethics & Self-Reflection

Curriculum Area: Ethics / Civic Engagement / Technology & Society
Subject / Strand: Discerning Screen Use for Research and Creativity
Lesson Title: AI Ethics & Self-Reflection (Adolescence 12–18)
Age Range: 12–18 years

Prerequisites / Previous Work:
Familiarity with AI systems, data collection, and the impact of technology on society. Builds upon earlier competencies in digital citizenship, design thinking, and bias awareness.

Materials:

- Case studies involving real-world AI ethical dilemmas (e.g., facial recognition, algorithmic bias, data privacy)
- Discussion prompts to provoke ethical reasoning and open-ended inquiry
- Role-play scenarios to explore different perspectives within ethical dilemmas
- Ethics frameworks: fairness, accountability, transparency, inclusivity
- Reflection journals or shared documents for writing and analysis

Presentation:

Prepare a respectful and focused discussion space with access to case studies and role-play materials. Invite the student and introduce the theme: "Today, we'll examine the moral questions that come with creating and using AI. What's fair? What's safe? What's right?" Begin by presenting a contemporary AI dilemma (e.g., the use of predictive policing or AI hiring tools). Guide the student in unpacking the scenario and identifying the core ethical issues. Facilitate a group discussion using prompts, then move into role-play—students choose roles such as developer, policymaker, or affected community member to deepen perspective. Close with a group reflection and a written or verbal summary of ethical takeaways.

Language:
Introduce and reinforce terms such as fairness, bias, accountability, transparency, harm, empathy, equity, informed consent, and ethical frameworks. Support students in applying these terms with specificity and integrity.

Points of Interest:
Students are deeply engaged when discussions center on fairness, safety, and inclusion. Ethical reflection provides a platform for voice, empathy, and respectful disagreement.

Direct Aims:
To develop critical thinking and ethical reasoning in the context of AI
To foster self-reflection and empathy through role-play and dialogue

Indirect Aims:
To prepare students for leadership in ethical innovation and civic participation
To build habits of inquiry, humility, and responsibility in tech-based decision-making

Control of Error:
Ethical discussions are open-ended but grounded in reflection. Gaps in logic, dismissiveness, or lack of clarity provide moments for guided questioning and deeper inquiry.

Extensions / Variations:
Students research historical ethical dilemmas in science and draw parallels to AI development. Facilitate student-led debates on emerging technologies. Guide students in designing AI projects with embedded ethical reviews or impact statements.

Follow-up Work / Assessment:
Observe how students participate in discussion and role-play. Note their ability to identify multiple sides of an issue and whether they demonstrate thoughtful self-reflection. Encourage them to journal about how their views evolved or deepened during the activity.

Notes for the Guide:
Ethics and self-reflection are the soul of AI education. Empowering adolescents to think beyond what technology *can* do, and toward what it *should* do, ensures a future shaped by fairness, foresight, and responsibility. These conversations are foundational to the moral development of creators, users, and leaders in a digitized world.

Chapter 58

Classroom Posters

These posters are designed to serve as quiet yet powerful invitations for conversation and reflection. Placed intentionally in Montessori classrooms, they highlight how core Montessori materials and practices naturally align with the skills needed in the Age of AI. Each poster offers a developmentally grounded talking point for visiting parents, guides, and stakeholders, sparking meaningful dialogue about how the Montessori method continues to prepare children for a rapidly changing world while remaining rooted in hands-on, human-centered learning:

Complete Set of Classroom Posters:
https://tinyurl.com/MappingPosters

Class Posters Summary

Mapping Montessori Materials for AI Competency Development

Class Posters

0-3

Mapping Montessori Materials for AI Competency Development

These competencies are acquired asynchronously, allowing students to develop and refine them at their own pace, based on individual readiness and engagement:

Social Intelligence & Empathy – Grace & courtesy lessons, observing facial expressions, mirroring emotions.

Order & Sequence Awareness – Arranging materials on a tray, following a simple routine, setting a table.

Spatial Awareness – Nesting cups, stacking blocks, crawling through tunnels, object permanence work.

Pattern Recognition – Matching colors, sorting by size, simple sound games, responsive caregiving.

Cause & Effect – Rattles, tactile mobiles, rolling a ball down a ramp, object permanence boxes.

Hand-Eye Coordination – Transferring snack with tongs, threading beads (infilare materials), placing pegs into a board (imbucare materials).

Tactile & Sensory Data Processing – Experiments in nature, texture matching, treasure baskets, mystery bag,

Simple Problem-solving Steps – Unlocking boxes, fitting shapes into a puzzle, trial-and-error activities.

Class Poster 0-3

Mapping Montessori Materials for AI Competency Development

Class Posters

3-6

Mapping Montessori Materials for AI Competency Development

These competencies are acquired asynchronously, allowing students to develop and refine them at their own pace, based on individual readiness and engagement:

Order & Symmetry – Arranging pink tower, knobbed cylinders, geometric cabinet, setting up workspaces.

Motor Skills & Coordination – Pouring grains, using tweezers, buttoning and zipping frames, clay modeling.

Algorithmic Thinking Foundations – Step-by-step tasks, moving objects in sequence, bead stairwork.

Observational Skills – Silence game, nature walks, object matching games, storytelling with visual cues.

Data Sorting & Classification – Sorting shells, leaf or animal cards, geometric shapes, classification charts.

Attention to Detail – Sandpaper letters, mystery bag, binomial cube, identifying small changes in materials.

Digital Citizenship as Practical Life – Grace & courtesy around devices, discussions on responsible choices, storytelling with picture sequences.

Class Poster 3-6

Mapping Montessori Materials for AI Competency Development

Class Posters

6-12

Mapping Montessori Materials for AI Competency Development

These competencies are acquired asynchronously, allowing students to develop and refine them at their own pace, based on individual readiness and engagement:

Precision & Control – Metal insets, cursive writing practice, bead frame math, scientific experiments.

Sequential Logic – Sentence analysis, long division work, timeline activities, coding unplugged games.

Basic Computational Thinking – Using decimal system materials, pattern extension exercises, storytelling with logic-based choices.

Conceptual Understanding of Patterns – Fibonacci spiral exploration, tessellations, geometry in art.

Basic Data Analysis – Graphing favorite activities, recording weather patterns, tracking plant growth.

Problem-solving with Complexity – Open-ended math problems, practical life troubleshooting, bridge-building challenges.

Algorithm Creation – Writing simple instructions for a friend to follow, map-making, coding board games.

Ethical Considerations – Group discussions on fairness, responsibility in digital spaces, moral dilemma stories.

Class Poster 6-12

Mapping Montessori Materials for AI Competency Development

Class Posters

12-18

Mapping Montessori Materials for AI Competency Development

These competencies are acquired asynchronously, allowing students to develop and refine them at their own pace, based on individual readiness and engagement:

Mathematical Foundations in AI – Probability exercises, matrices, statistics, real-world data applications.

Indirect Preparation for Natural Language Processing – Sentence diagramming, language deconstruction, poetry analysis.

Advanced Computational Thinking – Hands-on coding projects, logic puzzles, designing functional robots.

Data Structuring & Analysis – Spreadsheet work, cataloging library books, analyzing environmental impact data.

Data Visualization – Creating infographics, mapping historical events, using charts in research presentations.

Algorithm Optimization – Refining a process for efficiency, debugging code, improving an experiment's methodology.

Machine Learning Foundations – Training simple AI models (e.g., teaching an AI chatbot), game strategy analysis.

Project-based AI Applications – Creating AI-related research projects, building chatbots, AI in creative arts.

AI Ethics & Self-Reflection – Debates on AI's role in society, ethical case studies, personal reflection journals on technology use.

Class Poster 12-18

While these posters offer a visual entry point into the conversation, illuminating how Montessori materials support future-ready skills, they are just the beginning. To further support guides in observing and documenting children's growth in these areas, we offer a suggested rubric. Rooted in Montessori principles and inspired by reflective practice, this rubric is not a test or checklist, but a guide, a companion to help educators articulate, refine, and deepen their understanding of how AI-related competencies may already be emerging through the child's work.

To complement the rubric and deepen everyday usability, we have also created a simplified format for quick reference. Designed with busy classroom environments in mind, this version offers a streamlined way to connect Montessori materials with AI competencies at a glance, supporting guides in integrating this work fluidly into their daily observations and planning.

Chapter 59

Cheat Sheet #1: Montessori–AI Concepts

Simplified by Age & AI Competency

As well as the posters, this simplified version aligns with the AI competency categories used in this book. Again, it is grouped by age level and includes only Montessori terms and materials, the directly related AI competency, and a short classroom explanation. It is designed to be printed and displayed in classrooms or added to a teacher's binder. Once again, please remember this is a toe in the water exercise and your own observations will be vital as you develop your own mapping process.

Ages 0–3

Montessori Concept/Material	AI Competency	Simple Explanation
Routine & Order	Sequencing / Algorithms	Routines teach step-by-step thinking.
Object Permanence Box	Memory / Attention	Builds awareness of continuity.
Cause & Effect Play	Logic / Reasoning	If-this-then-that play mirrors AI logic.
Treasure Baskets	Data Collection	Exploring with senses = input gathering.
Hand-Eye Coordination	Robotics / Sensorimotor	Learning to guide motion and feedback.
Spatial Awareness	Mapping / Navigation	Moving through space is early pathfinding.

Montessori-AI Competency Cheat Sheet 0-3

Ages 3–6

Montessori Concept/Material	**AI Competency**	**Simple Explanation**
Pink Tower	Pattern Recognition	Builds size and order discrimination.
Color Tablets / Cylinders	Visual Classification	Sorting and matching develops categorization.
Sandpaper Letters	Language Processing	Connects symbols to sounds (early NLP).
Practical Life (Pouring, etc.)	Algorithmic Thinking	Tasks follow logical steps (like code).
Grace & Courtesy	Digital Citizenship	Kindness and turn-taking = safe tech use.
Binomial Cube	Abstraction / Problem Solving	Visualizes complex, layered patterns.
Nature Observation	Data Awareness	Watch, wonder, and compare over time.

Montessori-AI Competency Cheat Sheet 3-6

Ages 6–12

Montessori Concept/Material	AI Competency	Simple Explanation
Long Division Board / Checkerboard	Sequential Logic	Step-by-step cycles resemble loops.
Timeline of Life / Research	Data Analysis & Synthesis	Research projects build analysis skills.
Golden Beads (Large Numbers)	Data Representation	Concrete math shows abstract number systems.
Geometry Materials	Spatial Reasoning	Combining shapes = early spatial AI logic.
Advanced Practical Life	Precision & Calibration	Accuracy matters, like in programming.
Elementary Robotics / Coding	Algorithm Design	Children build and debug real instructions.
Grace & Courtesy in Tech Use	Digital Citizenship	Teaching responsibility with digital tools.

Montessori-AI Competency Cheat Sheet 6-12

Ages 12–18

Montessori Concept/Material	AI Competency	Simple Explanation
Algebra / Statistics Projects	Foundations of Machine Learning	Advanced math supports model building.
Literature & Grammar Analysis	Natural Language Processing	Breaks down sentence patterns like AI does.
Internship / Microeconomy	Systems Optimization	Learn how to adjust parts for better results.
Data Science Projects	Data Literacy & Interpretation	Collect, graph, and reflect on trends.
Coding / Programming Projects	AI Model Design	Students build real-world AI projects.
Ethics Discussions	Responsible AI Use	Explore fairness, privacy, and impact.

Montessori-AI Competency Cheat Sheet 12-18

Teacher Tip: These charts are for daily reference. They help connect your Montessori work to real-world, future-ready AI competencies, without adding technology. Let the child's work remain at the center and use this as a reflection guide. In the next chapter a second cheat sheet has more complex descriptions and examples. As you revisit this information presented in different ways, you will hopefully get a feel for how you and your group wish to handle this complex task.

As your familiarity with these ideas deepens, you may find value in seeing the connections presented with greater nuance. The next reference offers a broader view, expanding on the foundations with more layered examples and terminology. This evolving cheat sheet is meant to prompt deeper reflection, invite discussion, and affirm the meaningful overlap between Montessori practice and AI literacy.

Chapter 60

Cheat Sheet #2: Montessori–AI Concepts

Montessori–AI Concepts Cheat Sheet

This cheat sheet links further Montessori materials and concepts with AI/technology concepts at different developmental stages. Each age group highlights key Montessori terms or materials, the related AI concept, a simple explanation, and an example. These mappings show how hands-on Montessori learning naturally builds skills for the digital age. While not intended to be exhaustive, this evolving list serves as an invitation to deeper conversation within our schools and professional communities.

0–3 (Infants & Toddlers)

- **Order & Routine** – *Sequential Thinking (Algorithms)*:Infants and toddlers thrive on consistent routines (like the sequence of a bedtime ritual). This nurtures an understanding of order and step-by-step processes – much like a simple algorithm. *Example:* Following the daily routine (snack time after play time) mirrors how a computer follows a set sequence of instructions.

- **Cause & Effect Play** – *Basic Logic*: Babies repeatedly drop objects or push buttons to see what happens, learning that actions cause outcomes. This is the beginning of logical reasoning (if I do A,B happens), which underlies how computers use if/then logic. *Example:* A child flips a light switch (action) to turn a light on/off (result), similar to how pressing a key on a toy produces a sound.

- **Object Permanence Box** – *Memory & Continuity*: This Montessori material (a box where a ball disappears and reappears) teaches infants that things exist even when out of sight. It parallels how AI must remember information and track objects over time. *Example:* A baby drops a ball in the box and expects it to roll out again – akin to how a security camera AI remembers an object that temporarily goes off-screen.

- **Hand-Eye Coordination** – *Robotics & Sensorimotor Skills*: Activities like stacking rings or fitting shapes into a sorter help toddlers coordinate vision and movement. This resembles the sensorimotor integration a robot needs to physically interact with the world. *Example:* Placing a cylinder in the correct hole requires judging shapes and guiding the hand, much like a robot arm aligning a part correctly.

- **Spatial Awareness** – *Navigation & Mapping*: Crawling, climbing, and exploring space give toddlers a sense of distance, size, and position. This foundational spatial sense is similar to how AI in robotics or mapping software learns to navigate and understand environments. *Example:* A toddler learning not to bump into furniture is akin to a self-driving robot using sensors to avoid obstacles. Later in the 3-6 classroom the child will continue their spatial awareness explorations with the Pink Tower material.

- **Tactile Sensory Exploration** – *Data Collection (Sensors)*: Montessori infants use treasure baskets and sensory materials to explore textures, weights, and sounds. They are essentially "collecting data" through senses. Similarly, AI systems use sensors (cameras, microphones) to gather input about the world. *Example:* A toddler feeling rough sandpaper vs. smooth fabric is like a touch sensor detecting different surface textures.

- **Pattern Recognition** – *Finding Patterns*: Even in infancy, children recognize simple patterns (e.g. peek-a-boo timing or routine events). Montessori mobiles and rhythm games build on this. Recognizing patterns is exactly what AI does when it learns from data (like finding trends). *Example:* A baby anticipates that after "peek-a-boo" hands will open to a smiling face, similar to how an AI predicts the next item in a sequence after training on prior patterns.

- **Social Interaction & Empathy** – *AI Social Intelligence*: Toddlers learn social cues by observing faces and emotions, building empathy. While AI doesn't feel empathy, it can be trained to recognize emotional cues (like facial expressions). Montessori grace-and-courtesy lessons at this age (like gentle touching, taking turns) lay the groundwork for understanding emotional intelligence, an area of AI that involves interpreting human emotions. *Example:* A toddler comforting a friend who is crying demonstrates empathy; similarly, some social robots are programmed to detect sadness in a voice and respond with comforting words.

3–6 (Early Childhood)

- **Pink Tower & Brown Stair** – *Size Ordering (Data Sorting)*: These sensorial materials involve grading cubes or prisms by size, fostering visual discrimination and an understanding of seriation (ordering by a rule). This relates to how computers sort data from smallest to largest. *Example:* A child builds the Pink Tower from biggest cube to the smallest, much like an algorithm sorting numbers from highest to lowest.

- **Color Tablets & Knobbed Cylinders** – *Classification & Visual Pattern Recognition*: Children learn to sort color tablets by shade and fit cylinders into the correct holes by diameter. They are classifying and noticing subtle differences, mirroring how image recognition AI classifies objects by features (color, shape). *Example:* Matching pairs of color tablets (light to dark) is akin to an AI grouping images by color tone.

- **Practical Life Exercises (Pouring, Threading)** – *Algorithmic Thinking Foundations*: Activities like pouring water between pitchers, bead stringing, or handwashing have a clear sequence of steps. By practicing these sequences, children form mental "algorithms." They build concentration and logical ordering needed for coding later. *Example:* The four-step process of handwashing (roll sleeves, wet hands, soap, rinse) parallels a simple program with a defined sequence of actions.

- **Sandpaper Letters & Moveable Alphabet** – *Language Encoding (AI Language)*: Tracing sandpaper letters and forming words with the moveable alphabet teaches that symbols represent sounds and ideas, essentially data encoding. This links to how natural language processing AIs encode text into data. *Example:* A child sounding out "cat" with the moveable alphabet connects symbols to meaning, similar to how speech recognition software turns audio into text it can understand.

- **Observation of Nature** – *Scientific Inquiry (Data Observation)*: Montessori 3–6 encourages observing plants, animals, and weather (through nature tables or habitat puzzles). Children learn to notice and record details. This observational skill is akin to how AI gathers and analyzes data in scientific research. *Example:* A child watching and charting daily weather uses data collection and pattern spotting, which is the basis of data analysis in AI.

- **Binomial Cube (Pattern Puzzle)** – *Abstract Pattern Recognition*: The binomial cube is a 3D puzzle that concretely represents the algebraic pattern (a + b)3(a + b)^3(a + b)3. Children solve it by recognizing color and size patterns. This develops an ability to see patterns in abstract information, much like AI identifies patterns in complex data sets. *Example:* Solving the binomial cube by pattern cues is analogous to identifying a hidden rule in a data set (the way AI might, for instance, recognize a pattern in sales trends).

- **Attention to Detail (Control of Error)** – *Debugging and Error Correction*: Montessori materials often built in control of error (the material itself shows the child if something's wrong, e.g., a cylinder will not fit if in the wrong hole). This trains children to spot mistakes and correct them independently. In technology, this is like debugging code or spotting errors in data. *Example:* If one knobbed cylinder is left over after filling the block, the child realizes one hole has the wrong piece, similar to how a programmer finds which step of code produced a bug.

- **Grace & Courtesy Lessons** – *Digital Citizenship (Responsible Tech Use)*: Children at this age learn polite social behaviors (saying "please," taking turns speaking). Montessori educators map this to digital etiquette as an early form of digital citizenship. Just as they learn to be kind and respectful in person, children can begin to understand being safe and kind with technology (like asking before taking someone's picture, or gentle handling of devices). *Example:* A class role-play of politely asking before borrowing a toy translates to asking permission before using a tablet or sharing a photo. Montessori philosophy supports adding such *age-appropriate* tech courtesy lessons as part of Practical Life.

6–12 (Elementary)

- **Advanced Practical Life (Cooking, Woodworking)** – *Precision & Control (Calibration)*: Elementary projects like following a recipe or building a small woodworking project require precise measurement, planning, and control of tools. This parallels the precision needed in programming and engineering. *Example:* Measuring ingredients exactly to bake bread is like calibrating a scientific instrument or writing code with exact syntax, small errors can spoil the result.

- **Long Division Board & Checkerboard (Math Materials)** – *Sequential Logic*: Montessori math materials for long division or large multiplication guide children through multi-step operations in order. This instills an understanding of sequential logic (do step 1, then 2, then 3…) which is fundamental in coding and algorithms. *Example:* Using the long division board involves repeating a cycle of "divide-multiply-subtract" steps, much as a computer program runs a loop in sequence.

- **Hierarchy of Numbers (Golden Beads to Millions)** – *Place Value & Data Representation*: By working with golden bead materials from units to thousands to millions, children grasp base-ten representation and scaling. This is analogous to how computers use place value in binary (or any base) to represent data. *Example:* Composing 4,536 with beads (4 thousand,5 hundreds, 3 tens, 6 units) is similar to how a computer breaks a large number into bytes, a structured representation that AI algorithms manipulate.

- **Geometry Cabinet & Constructive Triangles** – *Pattern & Spatial Reasoning*: Exploring geometric shapes and combining triangles to form figures helps children see patterns and spatial relationships. This conceptual understanding of patterns connects directly to skills used in computer graphics and pattern recognition algorithms. *Example:* Noticing that two small triangles can form one big triangle is like recognizing a pattern that small parts can compose a larger structure, akin to how image AI recognizes a larger object made of smaller features.

- **Timeline of Life & Research Projects** – *Data Analysis & Synthesis*: In Montessori elementary, students study timelines (e.g. Timeline of Life) and undertake research projects where they gather information and draw conclusions. They learn to analyze data points (historical events, scientific facts) and see trends or cause-effect. This is a precursor to data analysis in AI, finding meaning in data. *Example:* Charting the rise and fall of a animal population on a graph and discussing why it changed is similar to an AI analyzing a dataset for trends and reasons.

- **Problem-Solving Activities (Going Out, Community Projects)** –*Complex Problem Solving*: Elementary students plan "Going Out" fieldtrips or classroom businesses (micro-economies), which involve solving real-world problems with many variables. They must be adaptable and think critically, just as AI tackles complex problems by breaking them down. *Example:* Organizing a bake sale to fund a charity involves budgeting, marketing, and teamwork, analogous to a complex project an AI might manage by dividing into subtasks (though humans handle the creativity and empathy parts!).

- **Elementary Coding & Robotics** – *Algorithm Creation*: Some Montessori programs introduce basic coding (e.g. Scratch programming or robotics kits) in late elementary. Students create their own simple programs or robot instructions, literally practicing algorithm creation. This hands-on coding relates Montessori's self-directed learning to real tech skills. *Example:* A 10-year-old uses a visual block programming app to make a robot follow a line, learning to create and adjust an algorithm (instructions) to achieve the goal.

- **Moral Discussions & Cosmic Education** – *Ethical Considerations*: Montessori elementary emphasizes cosmic education, understanding the larger human community and moral responsibility. In the context of AI, students can handle basic discussions about right and wrong uses of technology (fairness, honesty, impact on society). This cultivates an ethical mindset about technology. *Example:* A class discussion about whether using an AI chatbot to do homework is honest helps children consider AI ethics, paralleling how technologists must consider ethical use of AI systems.

12–18 (Adolescents)

- **Abstract Math (Algebra, Statistics)** – *Mathematical Foundations of AI*: Montessori adolescents delve into advanced math using concrete materials (like algebraic factor tiles or real-life data for statistics). Mastery of algebra, probability, and logic forms the backbone for understanding how AI algorithms function. *Example:* Working out a quadratic equation or analyzing survey data in class parallels the math behind machine learning models (which use equations and statistics to make predictions).

- **Language & Humanities Studies** – *Natural Language Processing (NLP)*: Adolescents study literature, composition, and maybe linguistics. They can connect this to how AI processes language. For instance, understanding grammar and sentence structure helps in grasping how chatbots or translation programs parse sentences. *Example:* A student diagramming a complex sentence is like an NLP algorithm breaking down a sentence to understand its grammar.

- **Computer Science Electives** – *Advanced Computational Thinking*: In a Montessori high school, students might have electives in programming, logic, or robotics. These experiences develop advanced computational thinking – solving problems with algorithms, using abstraction, and optimizing solutions. *Example:* Writing a Python program to simulate an ecosystem in science class involves decomposition of the problem and algorithmic thinking, just as real AI developers do when coding systems.

- **Data Science Projects** – *Data Structuring & Analysis*: Adolescents can undertake projects that involve collecting and analyzing data (e.g. a sociology survey or a science experiment tracking variables). They learn to organize data (spreadsheets, databases) and interpret results. This hands-on data literacy mirrors how data scientists train AI models by cleaning and analyzing datasets. *Example:* Students conducting a survey on school commute times and then graphing the results practice data analysis much like an AI finding patterns in user data.

- **Presentation & Graphics** – *Data Visualization*: Montessori values presenting research visually. Teens might create charts, infographics, or models to communicate findings. This correlates with data visualization in technology – presenting data in clear graphical form. *Example:* Creating an infographic on local water usage not only helps peers understand the info, but also mimics how AI outputs (like analytics dashboards) use charts to explain data.

- **Internship / Micro-Economy** – *Optimization & Systems Thinking*: Many Montessori adolescent programs include running a business (farm stand, café) or internships. Managing these requires optimizing processes (maximizing quality, efficiency), similar to how AI algorithms optimize solutions. Students learn to adjust variables to improve outcomes (e.g. best pricing, scheduling), a real-world parallel to fine-tuning an algorithm for better performance. *Example:* Teens running a school store analyze sales to decide optimal stock levels, akin to an AI system adjusting parameters to improve predictions.

- **Machine Learning Concepts** – *AI Model Building*: At the upper end, students may explore how AIs "learn" from data. They might use simplified machine learning tools (like training a simple model using an app or coding library) to see AI in action. This ties together math, coding, and data skills. *Example:* Using a teachable machine tool, students train an image classifier to recognize different leaves, experiencing how an AI model improves as it is fed more examples.

- **Community Service & Discussions** – *AI Ethics & Self-Reflection*: Montessori adolescents engage in community service and reflective seminars, examining their role in society. In terms of AI, they can discuss and debate the ethical issues of technology (privacy, bias, the impact on jobs) and reflect on how they use tech. This develops their moral compass and ability to foresee consequences, crucial for future AI innovators. *Example:* Students might debate a case study about facial recognition software's bias, exploring why diverse development teams and ethical guidelines are needed, which echoes Montessori's call for education that creates *ethical* future leaders in technology.

As you continue your own research and discoveries, we encourage you to share your insights generously. It is through collective integration, thoughtful dialogue, and ongoing refinement that our global Montessori community can co-create a unified mapping, one that is meaningful, inclusive, and accessible to all.

To support this shared work, we offer a sample rubric aligned with the values and structure familiar to Montessori educators. Grounded in observation and designed for flexibility, it aims to help guides recognize and record the emergence of AI-related competencies in a way that honors both the child and the method.

Chapter 61

Example of an AI Competency Rubric

This rubric is inspired by the *Montessori Assessment Playbook:* https://www.public-montessori.org/tools developed by the National Center for Montessori in the Public Sector (NCMPS). It offers a foundational structure for assessing AI-related competencies within Montessori environments, aligning with the Playbook's emphasis on observation, reflection, and developmental progression. Each competency is paired with indicators of understanding that emerge through authentic, hands-on work. Designed to be both adaptable and rooted in Montessori principles, this tool supports guides in documenting children's natural development in areas such as critical thinking, pattern recognition, ethical reasoning, and problem-solving. It serves as a reflective aid, not a standardized test, honoring the child's agency and the prepared environment. Montessori educators are encouraged to fine-tune and expand upon this rubric based on their unique classroom dynamics, ensuring that the integration of AI competencies remains child-centered and developmentally appropriate.

The inclusion of the NCMPS Playbook here is intentional. It represents a bridge, a living connection between the language we already know and use to support human development and the emerging language of competencies needed for the age of artificial intelligence. The Playbook's tools, such as the Normalization Check, Adjustment Check, Progress Report Templates, and Skills Inventories, are crafted from a deep understanding of Montessori pedagogy. They emphasize not only what a child knows but how a child grows. Their structure respects the wide scope of outcomes Montessori education is known for, executive functions, emotional flexibility, linguistic and cultural fluency, and social understanding.

Rather than reinventing the wheel, this rubric stands on the shoulders of these existing frameworks. It invites guides and school communities to bring their existing systems, whatever form they may take, into dialogue with a set of competencies informed by our changing world. The rubric does not seek to dictate a singular method, but rather to spark a shared project: the thoughtful construction of school-specific tools that reflect both Montessori values and future-readiness. In that spirit, schools are encouraged to develop their own rubrics, adapting the examples provided here or building upon the formats they already trust and use. The goal is not conformity, but conversation. And most importantly, it is collaboration.

As this work evolves, I imagine each school experimenting, refining, and discovering the language that best communicates what matters. What begins as a draft becomes a living tool, a shared vocabulary between child, guide, classroom, and community. We hope that when something resonates, educators will share it first within their teams and then within the broader Montessori movement. This is the kind of cooperative refinement Maria Montessori herself would have expected from a community of practice.

A note on the rating column: **"Level (E/P/M)"** refers to three developmental stages, **Emerging, Practicing, and Mastery.** These levels are used throughout the NCMPS inventories to acknowledge and respect the developmental arc of learning. An "Emerging" level reflects the first signs of a skill's development. The child is beginning to explore the competency and may need significant support. "Practicing" indicates increased consistency and growing independence. The child can perform the skill more frequently, often with minimal support. "Mastery" is not a final destination, but a reliable expression of deep understanding, independence, and integration. In the context of Montessori, this progression is not meant to measure the child against external benchmarks, but to help the adult prepare the environment in a way that supports the unfolding of the whole human being.

By including this familiar scale, we remain grounded in Montessori tradition while developing a common framework for discussing AI competencies across classrooms, age levels, and communities. In time, just as language for grace and courtesy or executive function emerged through shared observation, so too will a shared Montessori vocabulary around the competencies of the AI era.

This rubric, then, is more than a tool. It is an invitation, to observe deeply, to reflect honestly, and to build something together. Let us remember that no single rubric can capture the full essence of a child's growth. But when developed in community, rooted in shared purpose, and guided by love, these tools can help us prepare not just children for the world, but the world for children.

Here is a link to the whole Montessori & AI Mapping Rubric: https://tinyurl.com/Sample-Montessori-AI-Rubric

On the next few pages you will see visuals of the rubrics by level. Please remember, this is only an example and you will develop the best way to record this data for your school.

Mapping Montessori Materials for AI Competency Development: A Montessori Observation Rubric Informed by the NCMPS Assessment Playbook

(Levels (E/P/M) - Emerging, Practicing, Mastery)

School:	**Date/Year:**
Student:	
Class:	
Age:	

Age Group	Competency	Materials / Activities	Observation Criteria	Level (E/P/M)	Notes
0-3 Years	**Social, Intelligence & Empathy**	Mirror Play, Grace & Courtesy Lessons	Engages with peers respectfully, mirrors expressions, comforts others		
0-3 Years	**Order & Sequence Awareness**	Object Permanence Boxes, Daily Routines	Follows routines, sequences objects or actions correctly		
0-3 Years	**Spatial Awareness**	Nesting Boxes, Climbing Structures	Navigate space confidently, completes puzzles or fitting tasks		
0-3 Years	**Pattern Recognition**	Shape & Color Sorting, Sound Cylinders	Sorts by color, shape, or size; Identifies recurring patterns		
0-3 Years	**Cause & Effect**	Ball Ramp, Pop-Up Toys	Repeats actions to observe outcomes, predicts simple results		
0-3 Years	**Hand-Eye Coordination**	Pegs, Threading Beads	Transfers objects with accuracy, aligns actions		
0-3 Years	**Tactile & Sensory Data Processing**	Texture Boards, Sandpaper Letters	Explores and compares textures, responds to sensory input		
0-3 Years	**Simple Problem-Solving Steps**	Stacking Blocks, Two-Step Activities	Attempts solutions, shows persistence when challenged		
3-6 Years	**Order & Symmetry**	Pink Tower, Knobbed Cylinders	Arranges materials neatly, matches pairs or patterns symmetrically		
3-6 Years	**Motor Skills & Coordination**	Tweezers Work, Dressing Frames	Uses fine motor tools with precision, maintains posture		
3-6 Years	**Algorithmic Thinking Foundation**	Bead Stairs, Step-by-Step Cards	Follows multi-step tasks, notices when a step is skipped		

Sample AI Competency Rubric 0-3

Mapping Montessori Materials for AI Competency Development: A Montessori Observation Rubric Informed by the NCMPS Assessment Playbook

(Levels (E/P/M) - Emerging, Practicing, Mastery)

School: **Date/Year:**

Student:

Class:

Age:

Age Group	Competency	Materials / Activities	Observation Criteria	Level (E/P/M)	Notes
3-6 Years	**Order & Symmetry**	Pink Tower, Knobbed Cylinders	Arranges materials neatly, matches pairs or patterns symmetrically		
3-6 Years	**Motor Skills & Coordination**	Tweezers Work, Dressing Frames	Uses fine motor tools with precision, maintains posture during tasks		
3-6 Years	**Algorithmic Thinking Foundations**	Bead Stairs, Step-by-Step Cards	Follows multi-step tasks, notices when a step is skipped		
3-6 Years	**Observational Skills**	Silence Game, Nature Table	Describes objects, spots differences, focuses attention		
3-6 Years	**Data Sorting & Classification**	Zoology Cards, Leaf Cabinet	Group items by features, explains sorting logic		
3-6 Years	**Attention to Detail**	Sound Boxes, Color Tablets	Completes tasks without skipping steps, notices subtle differences		
3-6 Years	**Digital Citizenship as Practical Life**	Respect for Classroom Tech, Device Handling	Seeks permission, handles devices with care and intention		

Sample AI Competency Rubric 3-6

Mapping Montessori Materials for AI Competency Development: A Montessori Observation Rubric Informed by the NCMPS Assessment Playbook

(Levels (E/P/M) - Emerging, Practicing, Mastery)

School:	**Date/Year:**
Student:	
Class:	
Age:	

Age Group	Competency	Materials / Activities	Observation Criteria	Level (E/P/M)	Notes
6-12 Years	**Precision & Control**	Metal Insets, Geometry Cabinet	Applies pressure and motion deliberately, creates consistent output		
6-12 Years	**Sequential Logic**	Timeline of Life, Sentence Analysis	Arranges sequences accurately, explains logical progressions		
6-12 Years	**Basic Computational Thinking**	Golden Beads, Checkerboard Multiplication	Breaks problems into steps, applies patterns or strategies		
6-12 Years	**Conceptual Understanding of Patterns**	Fibonacci Spiral, Tessellations	Recognizes abstract similarities, creates visual or numeric sequences		
6-12 Years	**Basic Data Analysis**	Weather Charting, Survey Graphs	Collects information, draws conclusions from comparisons		
6-12 Years	**Problem-Solving with Complexity**	Word Problems, Science Experiments	Explores multiple solutions, revises ideas after testing		
6-12 Years	**Algorithm Creation**	Handmade Coding Cards, Logic Sequence Games	Designs clear multi-step instructions, tests and refines steps		
6-12 Years	**Ethical Considerations**	Class Debates, Moral Dilemmas Stories	Considers fairness, debates responsibility in group tasks		

Sample AI Competency Rubric 6-12

Mapping Montessori Materials for AI Competency Development: A Montessori Observation Rubric Informed by the NCMPS Assessment Playbook

(Levels (E/P/M) - Emerging, Practicing, Mastery)

School: **Date/Year:**

Student:

Class:

Age:

Age Group	Competency	Materials / Activities	Observation Criteria	Level (E/P/M)	Notes
12-18 Years	**Mathematical Foundations in AI**	Probability Work, Data Charts	Uses data to predict outcomes, applies logic to real scenarios		
12-18 Years	**Indirect Preparation for Natural Language Processing**	Sentence Diagramming, Poetry Study	Breaks down sentences, interprets meaning from word use		
12-18 Years	**Advanced Computational Thinking**	Scratch Programming, Grid Maps	Uses logic to solve unfamiliar problems, builds scalable solutions		
12-18 Years	**Data Structuring & Analysis**	Research Tables, Project Binders	Organizes information clearly, compares data sets meaningfully		
12-18 Years	**Data Visualization**	Infographics, Chart-Making Tools	Creates legible, attractive charts or visuals; explains them		
12-18 Years	**Algorithm Optimization**	Strategic Board Games, Coding Puzzles	Improves efficiency of steps or systems, tests alternates		
12-18 Years	**Machine Learning Foundations**	Pattern Prediction Activities, AI Simulations	Recognizes patterns, predicts outcomes, reflects on feedback		
12-18 Years	**Project-Based AI Applications**	Community Tech Projects, Capstone Research	Manages independent inquiry, applies tools toward purpose		
12-18 Years	**AI Ethics & Self-Reflection**	Ethics Journals, Tech Use Discussions	Reflects on tech's role, raises moral or social questions		

Sample AI Competency Rubric 12-18

Yet, even the most thoughtfully designed rubric is only as powerful as the spirit with which it is used. In Montessori education, it is not the framework but the guide who brings the learning to life, through observation, humility, and deep respect for the unfolding of each individual child. As we explore the development of AI literacy, we must return again to the spiritual preparation of the adult as the wise members of the IFTT technology group would frequently say. It is the prepared guide, centered and attuned, who ensures that every tool, whether tactile bead frame or digital prompt, is used with intention, reverence, and care. This next section honors that sacred role, reminding us that it is not only what we teach, but who we are, that shapes the future.

Chapter 62

The Spirit of the Guide in the Age of AI

Guiding children in ethical AI use is not merely a matter of technical competence; it is a call that begins within the heart of the adult. Dr. Maria Montessori urged the spiritual preparation of the teacher, a practice rooted in humility, inner reflection, and the ability to stand back and observe without prejudice. This preparation is not a single act but a lifelong commitment to cultivating awareness, presence, and respect for the unfolding development of each child.

Today, as we enter the age of AI, this inner work becomes ever more essential. The technologies shaping our world are not neutral tools, they are embedded with the values, intentions, and assumptions of their creators. As Montessori educators, we must pause to ask: What do we believe about AI? What unconscious biases do we bring to our understanding of technology? How is AI shaping the human experience, and how might it influence the development of children?

These questions do not require immediate answers. Rather, they ask us to enter a space of inquiry and wonder, a space Montessori knew well. In "The Absorbent Mind," Dr. Montessori reminds us that the adult must become a prepared observer, one who knows how to wait, who has faith in the child's ability to construct their own identity, and who remains humble before the mystery of human development. These same qualities are essential as we begin to engage with AI in our classrooms.

Ethical reflection in AI is not about teaching children a list of do's and don'ts. It is about guiding them toward a way of being that honors dignity, justice, and responsibility. It is about helping them see that technology is an extension of human will and that how we use it reflects who we are. It is about modeling the very qualities we hope to cultivate: discernment, compassion, curiosity, and courage.

To walk this path with integrity, we must rely not only on instinct or inspiration, but also on the depth of research traditions that have always guided Montessori practice. Dr. Montessori was, above all, a scientist, one who observed with rigor, questioned with curiosity, and constructed environments rooted in evidence and human development. As we seek to understand how to integrate AI thoughtfully and ethically, it is vital that we ground our inquiry in the very paradigms that shaped Montessori's work. The following chapter explores these foundational research paradigms, not as abstract theories, but as living tools to help us navigate this new frontier with discernment, grace, and a deep trust in the unfolding human potential.

Chapter 63

Montessori Research Paradigms as a Compass for AI Integration

Montessori education is grounded in research, observation, and the unfolding patterns of human development. Our pedagogy emerged from rigorous scientific study, not from ideology or reaction. Dr. Montessori's work is a model of inquiry-based innovation, guided by deep respect for the child and rooted in the real needs of society. This tradition offers us a steady compass as we navigate the integration of AI.

A research paradigm is a worldview, a lens through which we make sense of reality. In Montessori, these paradigms are lived every day through observation, reflection, and action. As we consider the possibilities of AI, five key Montessori-aligned paradigms can help ensure that our approach remains consistent with the child-centered values of the method:

Constructivism invites us to see AI not as a passive delivery system but as a medium for experimentation and creation. In the Montessori classroom, the child constructs their understanding through purposeful activity. AI can offer new avenues for this construction when used as a tool that supports agency, not automation.

Pragmatism reminds us that the true value of any educational tool lies in its usefulness. AI should not be adopted for its novelty but for its ability to meet real developmental needs. Montessori has always adapted to new contexts while remaining true to its core. The integration of AI must follow this same path: adaptive, responsive, and grounded in utility.

Humanism ensures that the child remains at the center. AI must never replace the relational, emotional, and moral dimensions of education. Instead, it can support children in developing the very qualities that make us human, empathy, creativity, and ethical reasoning. The human spirit, as Montessori taught us, is the force that shapes civilization. AI must serve this spirit.

Interpretivism values the uniqueness of each learner. While AI can offer personalized learning paths, it is the Montessori guide who interprets meaning, sees patterns, and connects learning to the child's lived experience. The guide remains the vital link between data and wisdom.

Naturalism affirms that real-world, hands-on experience remains essential. No screen can replace the sensory richness of the bead cabinet, the pink tower, or the garden. AI can extend the reach of the classroom, but it must never displace the child's engagement with nature, with materials, and with community.

Now is the time to take up this invitation. Let us, as Montessorians, researchers, and stewards of the child's future, come together to refine, share, and expand this work. Begin where you are, with one material, one classroom, one observation. Collaborate with your colleagues, adapt the mapping to your own school culture, and document what you learn. If something resonates, share it; if something does not fit, reshape it. This is not a finished product, it is a shared beginning. In honoring Dr. Montessori's legacy, we step forward as co-creators of an education prepared not just for today, but for the unfolding age of AI.

Chapter 64

A Call to Action: Co-Creating AI Literacy in the Montessori Community

We now arrive at a collective invitation. AI is not simply something to be implemented in the classroom; it is a new terrain that requires a co-created map. The Montessori guide, as observer, scientist, and prepared adult, is uniquely equipped to engage in this exploration with integrity and imagination.

Before diving into standards, rubrics, and competencies, we must acknowledge that this work is emergent. It will be shaped through cycles of observation, reflection, and iteration. Just as the child constructs the self through engagement with the environment, we must construct our understanding of AI through engagement with one another and with the unfolding realities of our classrooms.

Here are five interconnected pathways to co-create this work:

1. **Expand and Refine AI Competency Frameworks – L**et this book be a beginning, a placeholder for thought, not a fixed blueprint. Use the competencies provided to guide your inquiry, but allow them to evolve as you observe children in your unique context. Document what you see. Share your adaptations. Invite dialogue. Every observation adds richness to our collective understanding.

2. **Engage in Action Research –** Approach AI integration as a living study. Take notes on how children respond

to AI-related tasks. Reflect on whether the work fosters independence, ethical reasoning, or curiosity. Ask: Is this material aligned with Montessori principles? If not, how might it be adapted?

3. **Create a Montessori-Inspired Repository of AI Learning Materials** – We need more than just digital tools. We need materials that reflect the diversity, depth, and beauty of the Montessori classroom. Contribute new lessons. Translate competencies into tangible works. Ensure that what we build reflects the neurodiverse, culturally varied, and globally interconnected world our children inhabit and reflects their right to be seen, heard, and empowered.

4. **Champion Ethical AI Literacy and Digital Citizenship** – Guide children to use AI with care, humility, and discernment. Support them in asking questions: Who created this algorithm? What data was it trained on? What are the consequences of its use? Help them see that ethical AI use is not only possible but essential.

5. **Join the Montessori AI Literacy Movement** – Present your work. Share your questions. Contribute to a growing body of knowledge. Whether through conference presentations, research forums, or community blogs, your voice matters. Together, we can build a framework that is practical, principled, and rooted in Montessori wisdom.

And yet, as we take on this collective task, we must also pause. For even as we look outward, toward frameworks, technologies, and futures, we are reminded that true Montessori work begins within. Dr. Montessori taught us that the adult must prepare not only the environment, but themselves. In our observations of children, we see time and again that clarity of presence in the adult invites deeper concentration, peace, and purposeful work in the child. Neuroscience confirms what Montessori observed more than a century ago: that a regulated adult nervous system supports co-regulation in children, creating the conditions for learning, trust, and creativity.

Before we travel alongside children in the age of AI, we must clear space within ourselves, space for awareness, humility, and attentiveness. The art of clearing is not a luxury; it is a foundational practice for the Montessori guide, allowing us to meet the child, the moment, and the future with equanimity. It is here, in this stillness, that the deepest insights emerge. Let us turn now to that inner work.

Chapter 65

The Art of Clearing: Preparing the Inner Environment

Picture yourself standing on the playground. The sun is high, the chatter of children swirls around you, and suddenly a child barrels toward you, flushed and fuming, words spilling fast and urgent. Something has gone wrong, very wrong, and they need you to know.

“They were so unfair! It was horrible! I did not even do anything!”

You take a breath and meet them with calm. You listen, not just to the words, but to what’s underneath. And when there’s a pause, however brief, you gently ask, “Is there anything else?”

The question lands. It gives space.

And almost always, there *is* something else. A classroom moment from earlier that day, a look they misunderstood, a feeling they had not yet named. As they speak, their shoulders soften. Their voice settles. Eventually, they begin to see a fuller story, even their own role in what happened.

You ask again, “Is there anything else?”

And finally, a quiet, “No… I think that is it.”

You offer, “Would you like to stay with me a while? Would taking a walk help?” Sometimes they linger, grounded by your presence. But more often, they are off, lighter, freer, back to the swing, or the game, or even the very child they were upset with. And just like that, the moment has passed. That is clearing.

Children are remarkably good at it. They move through their emotions like weather, honest, brief, and real. They do not need to hold onto what they have felt; they simply feel it, speak it, and let it go.

We, as adults, struggle more. We carry old stories, old protections. We hold on. Internal Family Systems (IFS) offers language for this, it says that parts of us can get stuck, often at a very young age. A moment of feeling invisible at age four can harden into a belief that we are always alone. That we always have to manage everything ourselves.

But it does not have to stay that way. Montessori teaches us that transformation begins with the adult. When we clear, we become better prepared. When we model openness, curiosity, and grounded presence, we offer our students something far more powerful than a lesson, we offer them ourselves.

This is the art of clearing. And just like grace and courtesy, it is a practice. A way of being. A return to what is real.

Let us begin again.

The clearing practice described here did not originate with me. Its roots belong to Landmark Worldwide, where the idea of speaking what is present in order to return to clarity has long been taught and practiced. I encountered it as a practitioner and found it profoundly resonant with what Montessori guides already know to be true about the prepared adult. I offer it here in that spirit, with gratitude.

This is not therapy. But it is a chance to practice what children do so well. **Clearing.**

This simple exercise invites you to speak or write what is present, using the language of the stuck part, even if it sounds childish. You might say someone s being "mean" or "unfair." That is okay. The purpose is not to determine what is true. The purpose is to release what is inside so you can return to clarity.

Whether you are enamored by AI's potential or terrified by its implications, your opinions are not what serve the child. What serves them is your presence, your objectivity, your ability to follow their discoveries without inserting your own projections.

This Clearing Exercise, offered as a tool to help us all return to the clarity and lightness children show us every day. You can use it again and again, by yourself, or with others who understand the process, to move forward, to free yourself, and to stay in true service of the child.

As you continue mapping materials or introducing digital tools into the classroom, return to this exercise whenever you feel resistance, confusion, or uncertainty. It is in returning to presence that we remain true to our purpose.

May your path as a Montessori guide in the age of AI be one of clarity, courage, and care.

The Four Steps of the Clearing Exercise

1. What is the whole situation?
Begin by stating what is true, just three of four words, "The whole situation, is frustrating, scary, and exhausting." Say it out loud or write what is present for you in this moment, emotionally, mentally, or contextually, as you consider the role of AI in education. Do not worry about saying the good things, we do not need or want to clear those. Just the thoughts that are a hindrance to flow.

2. Who are the other people in this situation for you?
Reflect on those involved in your current situation, colleagues, parents, children, mentors, or public voices. Who do they represent to you emotionally or symbolically?

3. Who are you for others in this situation?
Consider how you imagine others perceive you. How are you showing up, skeptical, defensive, childish?

Do you feel a bit more clarity
Keep speaking until the feeling of a space emerges, notice if anything inside has softened or shifted.

4. What is the opportunity right now?
Finally, reflect on what becomes possible when you view this work with clarity. Words like, "Open, Joyful, and Peaceful" often arise, but it could be anything. Something that truly lights you up or excites you.

What might the children gain if you approach this work with openness and discernment?

Clearing Journal Format:

1. What is the whole situation?

2. Who are the other people in this situation for you

3. Who are you for others in this situation?

Keep speaking until the feeling of a mental space emerges.

4. What is the opportunity right now?

STAGES OF THE CLEARING EXERCISE

OVERALL SITUATION
Summarize the current environment

OTHER PEOPLE IN THE SITUATION
Identify thoughts about others involved

YOURSELF IN THE SITUATION
Identify thoughts about your own role

OTHERS' VIEW OF YOU
Identify thoughts others may have of you

OPPORTUNITY RIGHT NOW
Determine the present opportunity

From the spaciousness created through the clearing exercise, we now take a bold step forward, into territory that may feel less grounded, more complex. We are about to explore AI agents: digital tools capable of responding, interacting, even mimicking elements of our own thought and communication. I want to acknowledge that what you have just let go of in the clearing may now feel as if it is being filled back up again, but differently. If you have journeyed with me this far, I ask for just a little more of your presence, your patience, and your curiosity. What comes next is vital, not because AI agents are perfect or inevitable, but because how we choose to engage with them matters deeply. Let us walk forward, slowly and consciously, with clear eyes and open hearts.

Chapter 66

Montessori AI Agents & Student Agency

This book is not only about AI. It is not even only about Montessori. It is about agency, the deep, intrinsic force that lives inside every child and guides them to become a person of purpose, observation, and power.

Dr. Maria Montessori's greatest contribution to the world was not merely the creation of didactic materials or the architecture of a child-centered classroom. Her gift was understanding how to prepare an environment where agency could unfold naturally. At the 1915 Panama-Pacific International Exposition, she did not simply display children at work, she revealed that autonomy is not a lesson to be taught but a nature to be trusted. Her message was revolutionary: children are not empty vessels to be filled, but self-constructing beings, already endowed with purpose.

And now, we live in a world where agency, human agency, is under question once more.

In the age of artificial intelligence, it is easy to imagine that machines will outpace us, that systems will surpass our capacity to choose, reflect, or even participate in our own becoming. We are not introducing AI to children. We are recognizing that they are already digital citizens. They already live in a world of swipes and voice commands, of algorithmic influence and predictive text. They already know what it means to be shaped by unseen systems.

What they do not always know is that they can shape these systems in return. This is our moment to step forward, not as gatekeepers of the past, but as designers of the future. And in doing so, we offer our children the most Montessori gift of all: the invitation to create.

Imagine a Montessori environment where children do not merely use technology but create with it. Imagine a classroom where children build AI agents not as consumers but as curators of meaning, as researchers of relevance. Using tools like Chatbase, they can design an AI agent that answers questions about crickets or clouds, global warming or trust. But more importantly, they are learning that intelligence, even synthetic artificial intelligence, is only as meaningful as the sources, values, and stories we choose to give it.

Chatbots, Not Autonomous Agents

Chatbots

When we talk about introducing AI tools to children in a Montessori environment, it is important to distinguish between two very different kinds of systems. The first is what we will be focusing on together: question-and-answer style chatbots, such as those created in Chatbase or similar programs. These systems act like digital guides. The child builds the bot by feeding it information, organizing ideas, and shaping how it responds to questions. The bot does not act on its own; it simply answers based on what the learner has taught it. This approach aligns with Montessori principles because the child remains the active constructor of knowledge, engaged in classifying, experimenting, refining understanding, and seeing the relationship between input and output. Children practice careful thinking, digital responsibility, vocabulary development, and discernment as they decide what to include, how to phrase ideas, and how to guide a future user.

To create an AI agent is not to outsource thinking. It is to understand what thinking is. It is to learn what makes a source credible. It is to test how ideas connect. It is to see how language shapes action. And as Montessorians, we already know how to prepare for this. We do not rush to outcomes. We begin with observation. We follow the child. We protect concentration. We offer real tools. And we do not teach agency; we create the conditions for it to thrive.

Not, Autonomous Agents

By contrast, autonomous AI agents are tools that can take actions independently. They can make decisions, perform tasks across multiple steps, connect to external systems, and change things in the real world without ongoing human instruction. That level of automation can be powerful in professional environments, but it is not what we are cultivating in the classroom. For young learners, it is developmentally essential that thinking is visible, effort is experienced, and responsibility remains firmly with the child. Our goal is not to build automated systems that replace human judgment or initiative. Our purpose is to guide children in becoming thoughtful digital thinkers who design, question, and shape technology with clarity and ethical awareness.

In this way, AI becomes an extension of the prepared environment, introduced with the same respect, pacing, and guidance we offer with physical materials. It sits alongside hands-on work, never in place of it, supporting real work in the real world and strengthening the child's ability to communicate, sequence ideas, and reason with care. Understanding this distinction keeps AI positioned as a developmental material, introduced with intention and supported by the adult as a calm, observant guide. It remains a tool for strengthening thinking, creativity, and responsibility, not replacing the child's cognitive work or inner agency.

Creating AI agents can be introduced the same way we introduce any new material: through careful presentation, through respect for developmental readiness, and through the invitation to explore. At first, it may be collaborative, done in small groups, guided by a teacher. As interest grows, children will take it in their own direction. Some will want to make AI agents about nature, others about emotions, others about languages or sports or justice. Let them follow that thread.

These AI agents do not need to be published. They do not need to be shared. The learning is in the building. The conversation that unfolds between the child and their AI agent is the real material. And from this process, a new form of Practical Life emerges, not cleaning a table or pouring water, but designing the tools that shape tomorrow's knowledge.

We know, as Montessorians, that the child is always becoming. AI is not a threat to this becoming, it is a context in which that becoming now takes place. If we do not engage with it, we leave children to navigate it alone. But if we prepare the environment, if we offer the invitation, we find that the Montessori child is already ready.

Children do not need us to teach them how to use AI.
Children need us to show them that their voice belongs in its creation.

Let them build. Let them question. Let them imagine. Let them lead.

Because they are not behind the curve. They are the curve itself.

And so, with that truth ringing in our ears, that the children are not behind, but already ahead, we turn our attention to the practical: how we might begin to create with them. Not for them, but alongside them. The chapter that follows invites you into the process of building an AI agent, not as a technical exercise, but as a pedagogical one. A human one. This is where theory becomes practice, where curiosity becomes code, and where we, as Montessorians, remember that every tool, even the most advanced, is simply an extension of the child's inquiry. Let us step into this space not as experts, but as guides, ready to follow the child into the unknown, where their questions shape the future.

Chapter 67

Creating AI Agents in Schools: Child-Rights, Governance, and Environmental Consciousness

This short guide provides an overview of how educators and students can create an AI "agent" (a custom chatbot or assistant) using school-appropriate platforms. In K–12 settings, the central aim is not novelty. It is authorship. When students help build a constrained, inquiry-driven assistant grounded in teacher-curated materials, they move from passive consumption of AI outputs to active creation: shaping questions, selecting sources, testing claims, and reflecting on bias, limits, and ethics.

An AI agent can support student digital literacy, critical thinking, and research fluency. But it must be done in a way that protects children's rights, honors developmental needs, and meets school governance standards.

Learning Objectives

After engaging with this chapter, educators should be able to:

- explain what an AI agent is and how it functions within school-appropriate systems,
- distinguish between interacting with AI tools and authoring

constrained AI agents,

- evaluate AI platforms using child-rights, governance, and data-privacy criteria,
- identify developmentally appropriate thresholds for AI agent creation across age groups, and
- situate educational AI use within broader ethical and environmental systems.

Disclaimer

This guide is shared from professional experience and is intended as a resource to support educators exploring AI-agent creation in educational settings. The decision to proceed rests with each institution.

Educators are strongly encouraged to evaluate which AI platform best aligns with classroom age level, digital citizenship goals, institutional policies, and technical readiness. Before implementation, ensure your school or district has a clearly articulated AI and technology policy that includes guidance on the ethical use of generative AI, data retention, family communication, and student privacy.

All stakeholders, educators, administrators, students, and families, should be fully informed that this work is not about using AI to passively generate answers from third-party sources. Rather, it is about building constrained, purpose-built tools that reflect student-led inquiry and thoughtfully curated content.

Appropriate safeguards, privacy measures, and pedagogical alignment must be considered at every step. Seek collaboration with relevant colleagues (instructional technology, data privacy, special education, counseling, and IT/security teams) to ensure responsible and developmentally appropriate use.

Clear leadership accountability should be established. Final oversight for the selection, deployment, and review of AI tools rests with school leadership and, where applicable, district governance structures.

Developmental Readiness and Age Considerations

AI agent creation is not developmentally neutral. While older students may meaningfully engage in building, configuring, and testing AI agents, younger children should not be expected to do so.

In Montessori-informed environments, AI agent *creation* is most appropriate for upper elementary, adolescent, and secondary students who are developmentally capable of abstraction, systems thinking, and ethical reflection. Younger children may engage indirectly through observation, guided discussion, adult-modeled inquiry, and shared ethical conversations, without direct interaction or authorship.

Not all students need to build AI agents. There are many activities where the nuts and bolts of AI can be learned without any screens at all. In many classrooms, students may only interact with carefully prepared, adult-designed tools. Agent creation itself represents a higher-order competency and should be offered intentionally, not universally.

What an AI Agent Really Is

Most school-facing “agents” sit on top of a general large language model (LLM). Even without any school documents attached, an LLM can still produce fluent responses, summaries, and reasoning because it already contains broad prior training.

This is important to understand.

The AI is not empty. When a school “builds an agent,” it is not creating intelligence from zero. It is constraining and grounding an already-fluent system so it behaves like a specialist: aligned to a specific purpose, limited to approved materials, and governed by school rules.

In most education platforms, this happens through a pattern often described as retrieval + response (sometimes called retrieval-augmented generation).

The agent:

- receives a question,
- retrieves relevant passages from the approved knowledge set (documents, links, or a district repository),
- generates a response grounded in what it retrieved, and
- applies guardrails and policies set by the organization.

Enterprise agent platforms explicitly describe this concept as grounding agents in organizational data.

It is critical to understand that grounding limits *what the agent draws from*, not that it removes the underlying general intelligence. This is why careful testing, boundary-setting, and adult oversight are required.

School-Appropriate AI Agent Platforms Commonly Considered for District Approval

Because districts vary, "ISD-approved" is never universal. However, the platforms below are commonly approvable because they support the governance patterns districts typically require, including single sign-on (SSO), administrative controls, data protection agreements (DPAs), auditing, and documented student-data handling.

Microsoft Copilot Studio and Microsoft 365 Agent Tools (District Microsoft Tenants)

- For districts standardized on Microsoft 365, Copilot Studio is positioned as a platform to create, customize, and manage AI agents within enterprise governance structures.

- For organizations using SharePoint, Microsoft also supports agents that respond only to content a user already has permission to access, aligning with school access-control norms.

Google Cloud Vertex AI Agent Builder (District Google Cloud Governance)

- For districts with Google Cloud capacity and IT-managed infrastructure, Vertex AI Agent Builder is designed to build, scale, and govern enterprise-grade agents grounded in organizational data.

- This pathway is typically led by IT or innovation offices rather than individual classrooms, as it involves cloud projects, security configuration, and cost controls.

SchoolAI (Custom Student "Spaces")

- SchoolAI is widely discussed in K–12 contexts for creating student-facing, teacher-configured chat environments intended for guided exploration.

- The platform publishes FERPA and COPPA-related privacy documentation and district data-handling language. Local verification is always required.

MagicSchool (Custom Chatbot / Chat with Docs)

- MagicSchool offers custom chatbot and document-grounded chat tools designed for education use cases.
- The platform publishes privacy and security documentation describing how student data is processed under school agreements.

Platform inclusion in this list does not constitute endorsement. Final approval must follow district review processes.

How These Platforms Generally Work

Even though dashboards differ, most school-appropriate agent builders include the same functional components.

Purpose and Role Definition

You define what the agent is for (course assistant, research companion, "ask about this packet," bibliography helper, lab safety guide) and what it must not do (write essays, give final answers, provide medical advice, collect personal data). This is where pedagogical intent and developmental boundaries are established.

Knowledge Grounding

You attach the approved source set: teacher-curated documents, district policies, lesson materials, or a controlled repository. The agent is expected to reference only these materials when responding.

Guardrails and Safety Controls

You configure boundaries such as:

- staying within provided sources,
- refusal behaviors and escalation language,
- age-appropriate tone and content constraints,
- blocked topics
- citation requirements (where available),
- limits on memory, personalization, and user data collection.

Testing and Verification

You test the agent in a safe environment using control questions:

- before adding any sources, to observe default behavior,
- after adding sources, to verify grounding,
- and with boundary-testing prompts to identify hallucination or overreach.

Deployment and Access Control

You publish the agent inside approved channels such as district portals, LMS environments, or authenticated platforms, ideally with:

- single sign-on,
- role-based access,
- rostering where applicable,

- and audit logs.

Activity Review and Continuous Improvement

You review usage patterns not to surveil children, but to:

- detect confusion patterns,
- identify missing curriculum clarity,
- spot boundary failures,
- and refine prompts, sources, and safeguards.

AI agents should never be used as grading authorities, assessment tools, emotional supports, or decision-makers. They are instructional supports, not replacements for human judgment or relationship.

A Child-Rights Filter for Platform Selection

Before choosing any platform, evaluate it through a child-rights and school governance lens:

- Does it support a district DPA with clear subprocessor disclosures?
- Is student data used to train models by default, or explicitly excluded?
- Can the district control retention and deletion timelines?
- Can unnecessary logging, analytics, or identity capture be disabled?
- Is there SSO and role-based access?
- Are there educator controls to prevent the tool from doing the child's thinking?

For K–12 settings, these questions are foundational.

Family Partnership and Informed Consent

Families are not bystanders in this work. AI agent use should be communicated clearly to parents and caregivers in advance, including its purpose, boundaries, and data-handling practices.

Where required by law or policy, informed consent and opt-out pathways should be provided. Transparent communication builds trust and aligns home and school understanding around the responsible use of emerging technologies.

Conscious Use of AI Technology: The Physical Impact on World Resources

Artificial intelligence is not immaterial. Behind every agent sits physical infrastructure: data centers, electricity, cooling systems, and water.

Inviting students into AI creation also means inviting them into ecological awareness. This is not about alarm. It is about truthfulness and stewardship.

Classroom-scale AI use represents a small fraction of global AI energy consumption. However, making resource use visible supports systems thinking and ethical decision-making.

AI Agent Environmental Impact Tracker

Let us give students tools to see clearly. When they begin using AI agents, they can track the energy and water required. When they understand the environmental impact of their actions, they are empowered to make thoughtful decisions. Below is a sample table for students to use in real-time, as they create and engage with AI. It turns invisible systems visible: https://tinyurl.com/AI-Impact-Tracker

AI Agent Environmental Impact Tracker (Ages 6-18)

Assumptions for tracking: 0.5 kilowatt-hours of energy used per hour; approximately 1.8 liters of water used per kilowatt-hour. These numbers are averaged estimates based on published data and should be adjusted based on specific use cases or new research. (Please do your own research with your students before accepting these figures)

Date	Start Time	End Time	Duration (hrs./mins.)	Energy Used (kWh)	Water Used (liters)
2025-06-04	10:00 AM	10:30 AM	30 mins	0.25	0.45
2025-06-05	2:00 PM	2:45 PM	45 mins	0.375	0.675

AI Agent Usage Log: Environmental Impact Tracker

Assumptions for tracking:

0.5 kWh of energy use per hour (baseline placeholder; actual load varies widely).
Approximately 1–2 liters of water per kWh of electricity consumed (based on published WUE averages around 1.8 L/kWh). Figures should be adjusted based on regional infrastructure and cooling systems.

This table is not just a record. It is an invitation to consciousness. It is a way to pair the extraordinary creative power of AI with a deep sense of environmental stewardship.

Montessori Alignment: From Material to System

Creating an AI agent is not merely a technical exercise. It can become a living expression of competencies already mapped: observation, iteration, classification, precision, and ethical judgment.

Just as Montessori materials lead the child from concrete experience toward abstraction, AI agents can help older students reason about systems: where knowledge comes from, how evidence is evaluated, how language persuades, and how governance protects.

This work must remain consciously bounded. AI agents should not replace hands-on, sensory, real-world experiences. They should only complement them, preserving the child's agency, development, and contact with reality.

When prepared with care, AI agents can function as a new kind of material: responsive, constrained, and reflective, supporting inquiry without displacing the human guide or the lived world.

Reflection and Discussion Prompts

- In your educational context, where might AI agent creation clarify learning rather than complicate it?
- What developmental, ethical, or institutional boundaries would you establish before introducing AI agents to students?
- How does framing AI agents as constrained educational materials, rather than general-purpose tools, change your role as an educator or leader?

Chapter 68

Parents: Co-Guides in the Digital Age

The future of education is not shaped in the classroom alone. It is co-created by parents, educators, and the environments, both digital and physical, that surround the child. As artificial intelligence and advanced technologies become woven into the fabric of daily life, Montessori families and educators hold a shared responsibility: to help children develop digital literacy, ethical discernment, and the inner wisdom needed to navigate this evolving world. This guidance must remain deeply grounded in Montessori's hands-on, sensorial, and human-centered approach.

Dr. Maria Montessori often reminded us that the adult is not the center of the learning process, but a prepared observer and facilitator who supports the child's self-construction. Today, that role extends beyond the classroom to the home, where digital tools are first encountered and absorbed. The way technology is introduced at home shapes a child's understanding of its place in their life. Children need time, space, and thoughtful modeling to perceive technology not as entertainment, but as a tool for learning, expression, and responsibility.

In a Montessori home, digital literacy begins with curiosity. It does not begin with devices, but with questions. How does this voice assistant know what I said? What does a robot need to learn something new? Such inquiries open the door to conversations about logic, language, and ethics, conversations that are more valuable than any one app or program. When technology is introduced, it should be explored rather than consumed. Children benefit from opportunities to manipulate, construct, and experiment, with adult presence and guidance nearby but not interfering. Passive consumption creates dependency. Active exploration builds understanding.

These experiences must be anchored in the real. Montessori reminds us that sensorial exploration is the foundation for abstract thinking. Before a child learns about AI and pattern recognition, they should have touched pinecones, folded leaves, or arranged the Pink Tower with deliberate care. The physical world must come first, and always remain central. Screens are never substitutes for movement, nature, or human interaction.

In the same way we model grace and courtesy, we can now model conscious, ethical digital behavior. How we use our phones in front of children matters. How we speak about algorithms, privacy, or fairness shapes their inner world. The goal is not to shield children from technology, but to introduce it as we would any tool: with purpose, clarity, and respect. In this way, digital technology can become a practical life extension, something a child learns to use not for distraction, but for expression, creativity, and problem-solving.

None of this is possible without collaboration. Montessori education has always been a triangle, child, adult, and environment. Now, that triangle must extend across the threshold of school and home. Parents bring deep knowledge of their children and their family's values. Educators bring developmental understanding and the structure of the prepared environment. Together, they can ensure that digital technology is introduced intentionally, not reactively. They can create coherence for the child, so that the values cultivated in one space are reinforced in the other.

This is not a matter of keeping up with trends or fearing the future. It is a matter of stewardship. Our children are entering a world shaped by algorithms and automation, but they do not have to be shaped by them. With loving guidance, grounded in Montessori's enduring insights, they can become ethical, empowered, and deeply human learners, ready to lead with integrity in an AI-driven world. A future-ready pedagogy does not chase technology, it centers the child. It recognizes that the competencies needed in this new era, curiosity, discernment, collaboration, and care, are not separate from Montessori, but already alive within it. Our task is to bring these forward with intention, preparing not only the environment, but also ourselves, for a future where education remains a place of liberation, not limitation.

Chapter 69

A Future-Forward Pedagogy

The initiative to map Montessori materials to AI competency development deepens our collective understanding of how the Montessori method inherently fosters the cognitive, social, and problem-solving abilities most essential for the age of artificial intelligence. By systematically aligning Montessori materials with emerging AI and digital literacy competencies, this work provides early validation that Montessori pedagogy is not only developmentally sound but also future-relevant.

This mapping does not change the essence of Montessori, it reveals it more clearly in the context of 21st-century educational discourse. It offers a language for communicating Montessori's value in a world increasingly shaped by algorithms, automation, and complex systems thinking. Through this lens, Montessori is not outdated, it is foundational.

This chapter outlines the expected impact of the mapping process on the Montessori ecosystem, including teacher training, curriculum refinement, scientific research, and advocacy. These efforts are designed not to standardize Montessori practice, but to strengthen its voice in contemporary education.

Enhancing Montessori Teacher Training and Awareness – The AI competency mapping framework can be integrated into Montessori teacher preparation and professional development programs. It equips educators with new ways to observe and articulate the skills children are developing, particularly those connected to systems thinking, ethical reasoning, and self-direction. This clarity empowers guides to communicate Montessori's relevance to parents, policymakers, and interdisciplinary collaborators in education and technology.

Advocating for Montessori in 21st-Century Education – Mapping Montessori materials to AI-relevant competencies helps position Montessori as a future-ready pedagogy. It demonstrates how long-standing materials already foster pattern recognition, problem-solving, adaptability, and foundational computational thinking. These insights offer a compelling narrative for policymakers and school leaders, providing a clear bridge between timeless Montessori values and the urgent needs of modern learners.

Moreover, this work supports increased parental awareness. It helps families understand that Montessori is not just a philosophical or alternative model for early childhood, it is a deeply relevant, scientifically aligned approach that prepares children to lead in a rapidly changing, ethically complex world.

Refining Montessori Curriculum and Material Development – The mapping process also opens new opportunities for curriculum enhancement. By highlighting areas of strength and areas for growth, it supports thoughtful refinement of lessons and materials. For example, the mapping may reveal competencies that are well-supported in early childhood but less explicitly addressed in adolescence. This can lead to the creation of new extensions, interdisciplinary connections, or digital practical life activities that align with Montessori principles.

Additionally, material developers may use this framework to optimize the scope and sequence of resources across developmental planes, ensuring that competencies such as data analysis, ethical AI reasoning, and systems thinking are scaffolded throughout the curriculum without compromising hands-on, sensorial learning.

These goals are deeply aligned with global trends in AI and education:

According to **McKinsey & Company** (2023), up to 40% of a teacher's time could be redirected from routine administrative tasks through AI support, freeing them to focus more on direct student engagement and individualized learning (McKinsey & Company, 2023).

The **World Economic Forum** (2024) emphasizes that AI should be harnessed to personalize learning, promote equity, and enhance both pedagogy and assessment—provided it is implemented collaboratively and ethically (World Economic Forum, 2024).

Meanwhile, **UNESCO** (2023) advocates for a human-centered, ethical framework that supports age-appropriate AI integration, professional capacity-building, and policy development rooted in educational inclusion and justice (UNESCO, 2023).

These findings affirm the relevance and timeliness of the Montessori–AI mapping initiative. By rooting this work in research-based global frameworks, the Montessori community not only joins the AI education conversation, it leads it. At a time when many educational models are scrambling to retrofit their practices for a rapidly evolving technological landscape, Montessori offers something rare and essential: a method already grounded in observation, agency, and developmental wisdom.

Guiding Montessori into the age of AI is not about abandoning tradition. It is about reactivating its most powerful elements, its trust in the child, its reverence for the prepared environment, and its insistence that education is, above all, a moral and spiritual endeavor. In doing so, we ensure that technology remains a tool, not a tyrant; that AI is something our children shape with clarity and compassion, rather than something that shapes them.

Let this be our collective invitation to walk boldly into the future, carrying with us the timeless principles that have always defined Montessori education. The world does not need a new pedagogy to meet this moment. It needs a renewed confidence in the one that already knows how.

Chapter 70

Conclusion: Guiding Montessori into the Age of AI

Each Montessori environment must discern for itself whether, and how, to engage with artificial intelligence or create AI agents with students. For those who choose to explore its potential, the approach must be intentional, rooted in observation, and aligned with Montessori's enduring values. AI must serve as a complement, never a replacement, for the child's natural, hands-on learning journey.

Just as Montessori environments are carefully prepared, so too must any use of AI be thoughtful and developmentally appropriate and adapted to the unique needs of each child and classroom. It should support creativity, inquiry, and ethical reasoning without compromising the sensorial, experiential core of the Montessori method. A child who first explores math through golden beads may one day use AI to analyze data, but it is the tangible, self-directed foundation that makes such abstract thinking possible.

If integrated, AI can become an extension of scientific inquiry, an opportunity for students to observe, question, and experiment. This invites ethical reflection and digital citizenship into the curriculum in ways that are authentic to Montessori's peace education and social responsibility.

Used judiciously, AI may also support differentiated learning and creative expression, adapting to children's needs while amplifying their voices. Yet caution must remain: AI is a tool, not a substitute for the guide's wisdom or the child's originality.

There is no mandate to adopt or reject AI, but rather an invitation to engage with it as we do all things in Montessori: through observation, experimentation, and reverence for the child's unfolding potential. The mapping process does not introduce something new, it reveals the future-ready power of what already exists.

By naming age-related competencies and anchoring them in real-world relevance, we now have the language to engage stakeholders in meaningful dialogue. Montessori guides are the gatekeepers of physical reality, and the protectors of presence in an increasingly digital world. In an age of increasing abstraction, our commitment to nature-first, hands-on education anchors children in truth, beauty, and belonging.

This is a moment for the Montessori community to lead, not by chasing trends, but by demonstrating how timeless pedagogy meets the needs of a changing world. As Dr. Montessori reminded us:

"An education capable of saving humanity is no small undertaking; it involves the spiritual development of man, the enhancement of his value as an individual, and the preparation of young people to understand the times in which they live."

With this spirit in mind, the pages that follow are offered as a gentle bridge for all who wish to walk alongside this work, whether seasoned Montessorians, curious educators from other traditions, or anyone exploring what it means to raise children in the age of AI. You will find practical suggestions for locating Montessori materials, a brief note from the author, acknowledgments that center gratitude over individual credit, and a list of references that capture the beating heart of this work. Two glossaries are also included, offering definitions of both AI competencies and Montessori terms, to ensure that no one is left out of the conversation. This work is truly for everyone committed to nurturing wise, empowered children in a world yet to be made. Let us move forward together.

Chapter 71

Montessori Material Providers

This list is not a promotion or recommendation of these companies; it is solely for informational purposes. It is provided here to support the reader in understanding the materials commonly used in Montessori environments. The inclusion of a company on this list does not imply endorsement, nor does it guarantee the quality, authenticity, or compliance of their products with Montessori principles. Educators and schools are encouraged to conduct their own research and due diligence when selecting materials for their classrooms. For the non-Montessori reader, this list offers an opportunity to review the products described in the book.

The Montessori materials market is diverse, ranging from well-established manufacturers to independent craftspeople and digital resources. Below is an example of Montessori material suppliers, arranged alphabetically, there are many more:

- Affordable Montessori Material
- Alison's Montessori
- Bruins Montessori USA
- Gonzagarredi Montessori (GAM)
- Matsumoto Kagaku (Japan)
- Montessori Concepts

- Montessori-n-Such
- Montessori Outlet
- Montessori Services
- Nienhuis Montessori USA
- Pink Montessori
- The Materials Company of Boston

Chapter 72

Acknowledgments

This work was created during a time of profound transformation, both in education and in the world at large. It reflects the dynamic intersection of artificial intelligence, digital literacy, and Montessori pedagogy, all coming together to reimagine how we prepare children for the future.

In an era of rapid technological change, Montessori education continues to adapt while remaining grounded in its timeless principles. The formation of the American Montessori Society Innovation Fellowship Think Tank (2023–2025), and the collaborative work of the Technology Group who developed the groundbreaking exhibit *Innovation Labs: The Future Montessori Classroom* for the 65th Anniversary of The Montessori Event in Denver, March 2025, exemplify this thoughtful integration of innovation and tradition.

This mapping research was shaped by many meaningful conversations, generous guidance, and timely exploration. To every member of this process, thank you. Your insight, commitment, and curiosity have helped shape a vision that honors Montessori's enduring legacy while opening new pathways for future generations.

AMS IFTT technology group members (2023-2025): Debbie Adams, Dana Anderson, Pat Barton, Kate Broughton, Samantha Gleisten, Aishwarya Saminathan, Shelley Smith, Aderonke Okesola, Katie Torres, Michelle Velho, Mara Weitzman, Jovan Well.

AMS IFTT facilitators and support members: Hannah Baynham, Dean Blasé, Gabrielle Holt, Gina Lofquist, Rosemary Quaranta, Eric Rodriguez.

This document was created over the period of two years of notetaking and reflection. It is still a working document, and I am deeply grateful for the hands and minds who have helped this on its journey.

With deepest love and gratitude to my family, Mark, Amelia, Olivia, & Alex, whose faith, brilliance, and unwavering support have created the conditions for this work to take form. Their quiet strength, joyful presence, and generous spirit have made space for inquiry, for creative risk, and for wholehearted service.

This offering, crafted in service to educators everywhere, emerges not from one voice alone, but from the nourishment of a family rooted in trust and shared purpose. May the love that surrounds this work be felt in every page, every classroom, and every act of guidance offered to a child.

Chapter 73

About the Author

Kate Broughton, M.Ed., is an educator, artist, and practitioner-researcher with more than forty years of experience across a wide range of educational settings in the United Kingdom and the United States. Her work spans public education, private Montessori institutions, teacher preparation programs, and nonprofit school leadership, with a sustained focus on child development, embodied learning, and educational innovation.

Kate holds American Montessori Society (AMS) Elementary I & II and Administrator credentials, as well as a Master of Education in Montessori Integrative Learning from TIES in partnership with Endicott College. In 2007, she co-founded Kingwood Montessori School, a nonprofit Montessori institution in Texas, where she later served as director and as a member of the Board of Trustees. Her leadership contributed to the development of a learning environment grounded in Montessori principles and informed by contemporary research.

She has contributed to Montessori teacher education as faculty at the Gulf Coast Montessori Teacher Education Center and the Montessori Education Center of the Rockies and has engaged in curriculum and assessment work internationally. In 2023, Kate joined the inaugural cohort of the AMS Innovation Fellowship Think Tank, completing the two-year fellowship in 2025. Her award-winning research, *Mapping Montessori Materials for AI Competency Development*, explores how Montessori learning environments cultivate competencies relevant to emerging technologies while preserving human agency, judgment, and creativity.

Beyond Montessori, Kate has served advisory roles focused on leadership, strategy, and the responsible integration of artificial intelligence in education and organizational contexts. She has contributed to the Strategic AI program created by Z-School, hosted at Mercy University and the University of San Francisco, and has advised initiatives exploring the intersection of human-centered values, technology, and the future of work. This work reflects her commitment to ensuring that advances in AI are guided by developmental insight, ethical clarity, and long-term societal well-being.

Kate's background in the arts continues to inform her of an interdisciplinary approach to education. She holds a BA (Hons) degree in Fine Arts, was a Licentiateship with the Society of Designer Craftsmen (1981) and maintains an active creative practice that bridges traditional craftsmanship with emerging technologies.

She is the author of *Montessori & AI* (Volumes I and II), *Mapping Montessori Materials for AI Competency Development*, *After Alpha*, and *Create Montessori*. Her writing engages educators, leaders, and families seeking clarity, coherence, and confidence in a time of rapid technological change.

She currently lives and works in Colorado.

Chapter 74
References

AI4K12. (2022). *Artificial intelligence for K–12 initiative: Five big ideas of AI.*

Adadi, A., & Berrada, M. (2018). Peeking inside the black box: A survey on explainable artificial intelligence (XAI). *IEEE Access, 6*, 52138–52160.

American Montessori Society. (2023). *AMS Innovation Fellowship: Research initiatives for a changing world.*

American Montessori Society. (n.d.). *Core components of Montessori education.*

Association Montessori Internationale. (n.d.). *Cosmic education.*

Barocas, S., & Selbst, A. D. (2016). Big data's disparate impact. *California Law Review, 104*(3), 671–732.

Bostrom, N. (2014). *Superintelligence: Paths, dangers, strategies.* Oxford University Press.

Campbell, M., Hoane, A. J., & Hsu, F. H. (2002). Deep Blue. *Artificial Intelligence, 134*(1–2), 57–83.

Choudary, S. P. (2025). *Reshuffle: Who wins when AI restacks the knowledge economy*. Platformation Labs.

Crevier, D. (1993). *AI: The tumultuous history of the search for artificial intelligence*. Basic Books.

Dondi, M., Klier, J., Panier, F., & Schubert, J. (2021, June 25). *Defining the skills citizens will need in the future world of work*. McKinsey & Company.

Gerber, M. (2002). *Your self-confident baby: How to encourage your child's natural abilities—from the very start*. Wiley.

Gettman, D. (1987). *Basic Montessori: Learning activities for under-fives*. St. Martin's Press.

Goertzel, B. (2014). Artificial general intelligence: Concept, state of the art, and future prospects. *Journal of Artificial General Intelligence, 5*(1), 1–48.

Grazzini, C. (1996). The four planes of development. *The NAMTA Journal, 21*(2), 208–241.

Hainstock, E. G. (1997). *Teaching Montessori in the home: The preschool years*. Plume.

Helfrich, M. S. (2018). *Montessori learning in the 21st century: A guide for parents & teachers*. Parent Child Press.

Hodges, C., Moore, S., Lockee, B., Trust, T., & Bond, A. (2020). *The difference between emergency remote teaching and online learning. Educause Review*.

Indiana University. (n.d.). *Reweaving meaning in the Navajo spirit pathway. Museum Anthropology Review*.

Japan Objects. (2022, October 31). *What is wabi-sabi?*

Jones, R. (2015). *Descartes' bones: A skeletal history of the conflict between faith and reason*. W. W. Norton & Company.

Kasparov, G. (2017). *Deep thinking: Where machine intelligence ends and human creativity begins*. PublicAffairs.

Kramer, R. (1976). *Maria Montessori: A biography*. Putnam.

Kramer, R. (1988). *Maria Montessori: A biography*. University of Chicago Press.

Landmark Worldwide. (1991). *The Landmark Forum* [Personal and professional development program; clearing practice adapted for Montessori educational contexts]. Landmark Worldwide LLC.

LeCun, Y., Bengio, Y., & Hinton, G. (2015). Deep learning. *Nature, 521*(7553), 436–444.

Lillard, A. S. (2005). *Montessori: The science behind the genius*. Oxford University Press.

Lillard, A. S. (2013). Playful learning and Montessori education. *American Journal of Play*.

Lillard, A. S. (2017). *Montessori: The science behind the genius* (3rd ed.). Oxford University Press.

McCarthy, J., Minsky, M., Rochester, N., & Shannon, C. (1956). A proposal for the Dartmouth Summer Research Project on Artificial Intelligence. *AI Magazine, 27*(4), 12–14.

McCorduck, P. (2004). *Machines who think* (2nd ed.). A. K. Peters.

McKinsey & Company. (2023). *How artificial intelligence will impact K–12 teachers*.

McTamaney, C. (2009). *A delicate task: Teaching and learning on a Montessori path*. Parent Child Press.

Merriam-Webster. (n.d.). *Artificial intelligence*. In *Merriam-Webster.com dictionary*.

Montessori for Everyone. (n.d.). *The five great lessons*.

Montessori International College. (2019). *What are the great lessons?*

Montessori, M. (1913). *Pedagogical anthropology* (F. T. Cooper,

Trans.). Frederick A. Stokes Company. (Original work published 1913)

Montessori, M. (1964). *The Montessori method*. Schocken Books. (Original work published 1912)

Montessori, M. (1965). *The advanced Montessori method* (Vol. 1). Robert Bentley. (Original work published 1917)

Montessori, M. (1967). *The absorbent mind*. Holt, Rinehart and Winston. (Original work published 1949)

Montessori, M. (1972a). *Dr. Montessori's own handbook: A short guide to her ideas and materials*. Schocken Books.

Montessori, M. (1972b). *The secret of childhood*. Ballantine Books.

Montessori, M. (1989). *The child in the family*. Clio Press.

Montessori, M. (1989). *The formation of man*. Clio Press. (Original work published 1949)

Montessori, M. (1995). *Education and peace* (H. R. Lane, Trans.). Clio Press. (Original work published 1949)

Montessori, M. (1996). *From childhood to adolescence, including "Erdkinder" and the functions of the university*. Montessori-Pierson Publishing.

Montessori, M. (1997). *To educate the human potential*. Clio Press. (Original work published 1948)

Montessori, M. (2007). *To educate the human potential*. Montessori-Pierson Publishing.

Montessori, M. (2012). *The 1946 London lectures*. Montessori-Pierson Publishing. (Original lectures delivered 1946)

Montessori, M. (2007). *The discovery of the child*. Aakar Books.

(Original work published 1949)

Montessori, M., & McDermott, J. J. (1965). *Spontaneous activity in education*. Schocken Books.

Montessori-Pierson Publishing Company. (n.d.). *Collected works of Maria Montessori*.

NAMTA. (1990). *Montessori under three* (Special issue). *The NAMTA Journal, 15*(3).

NAMTA. (2001). *The Montessori adolescent: Analysis in retrospect* (Special issue). *The NAMTA Journal, 26*(3).

Nazmiyal Antique Rugs. (n.d.). *Old rugs and their imperfections: Understanding intentional flaws in antique rugs*.

North American Montessori Center. (2018, September 26). *What's the rush? When presenting the five great lessons*.

Organisation for Economic Co-operation and Development. (2023). *AI literacy for students: Framework for secondary education*.

Psychology Today. (2022, July 8). *Wabi-sabi and the psychology of imperfection*.

Richardson, M. (1953). *Art and the child*. University of London Press.

Russell, S. J., & Norvig, P. (2021). *Artificial intelligence: A modern approach* (4th ed.). Pearson.

Saminathan, A. (2025). *The sixth great lesson: The coming of technology*. American Montessori Society.

Skeptics Stack Exchange. (2017, May 2). *Did Persian rug makers traditionally include deliberate imperfections?*

Society of Designer Craftsmen. (n.d.). *Society of Designer Craftsmen*.

Standing, E. M. (1957). *Maria Montessori: Her life and work*. Plume.

Schwab, K. (2017). *The fourth industrial revolution*. Portfolio.

Stephenson, S. M. (2010). *The joyful child*. Michael Olaf.

Stephenson, S. M. (2013). *Child of the world*. Michael Olaf.

UNESCO. (2023). *Guidance for AI and education: A curriculum framework for AI literacy.*

UNESCO. (2024). *AI competency framework for students.*

University of Michigan, Kelsey Museum of Archaeology. (n.d.). *Less than perfect: Navajo spirit lines.*

Wing, J. M. (2006). Computational thinking. *Communications of the ACM, 49*(3), 33–35.

World Economic Forum. (2020). *Schools of the future: Defining new models of education for the fourth industrial revolution.*

World Economic Forum. (2024). *Shaping the future of learning: The role of AI in education 4.0.*

Zimmer, B., & Zimmer, J. (2007–2023). *Foundational support for Kingwood Montessori School*. Zimmer Foundation.

Chapter 75

Appendix A: Montessori Glossary Aligned with AI Competencies and Extensions

This glossary provides general definitions for foundational Montessori terms and aligns them with key AI-related competencies needed in the modern world. The purpose is to demonstrate how timeless Montessori principles naturally support the development of skills identified as critical for the Age of Artificial Intelligence. The AI competencies referenced here are based on internationally recognized frameworks from McKinsey & Company, UNESCO, and the World Economic Forum.

Each alignment reflects the spirit of Montessori education, one that respects the developmental arc of the child, encourages independence, and cultivates a prepared mind capable of ethical, abstract, and systems-level thinking.

Montessori Term | Definition | AI Competency Alignment

Absorbent Mind
A formative stage where the child's mind absorbs experiences with ease.
Curiosity, Cognitive Flexibility, Self-directed Learning

Classification Systems
Organizing information to construct mental order.

Pattern Recognition, Systems Thinking, Organization

Classified Nomenclature
Accurate naming of objects and ideas.
Information Literacy, Data Labeling, Vocabulary Development

Concrete to Abstract
Moving from tangible experiences to conceptual understanding.
Abstraction, Symbolic Reasoning, Conceptual Thinking

Control of Error
Self-correction embedded into materials.
Autonomous Problem Solving, Feedback Processing, Self-regulation

Control/Working Charts
Reference tools for self-checking operations.
Self-assessment, Error Correction, Confidence in Computation

Cosmic Task
A philosophical concept of purpose and contribution.
Ethical Reasoning, Systems Awareness, Purpose-driven Learning

Etymology
Study of word origins and evolution.
Linguistic Reasoning, Historical Analysis, Semantic Understanding

Follow the Child
Education directed by observation and individual developmental needs.
Adaptive Learning, Human-Centered Design, Personalization

Freedom Within Limits
Autonomy supported by clear boundaries and responsibility.
Responsible Agentic Behavior, Governance, Ethical Decision-making

Grace and Courtesy
Explicit teaching of empathy, civility, and respectful interaction.
Digital Citizenship, Social-Emotional Intelligence, Prosocial Online Conduct

Interrelatedness
Understanding that everything is connected.
Systems Thinking, Interdisciplinary Integration, Network Understanding

Isolation of Difficulty
Focusing on one skill at a time.
Sequencing, Targeted Skill Acquisition, Learning Design

Maximum Effort
Children seek meaningful physical and cognitive challenge.
Intrinsic Motivation, Effortful Learning, Executive Function Strengthening

Normalization (Concentration)
Emergence of deep focus, self-discipline, and joyful work.
Focused Attention, Self-regulation, Executive Function, Flow State Development

Practical Life
Real tasks that cultivate autonomy, coordination, and contribution.
Real-World Problem Solving, Embodied Cognition, Tool Fluency

Prepared Environment
Deliberately designed surroundings supporting independence, concentration, and real-world engagement.
Environment Design, Adaptive Learning Systems, Human-Centered Technology

Repetition
Children repeat activities until inner mastery is achieved.
Deliberate Practice, Mastery Learning, Motivation

Sensorial Materials
Tools to refine perception and classification of experience.
Perceptual Training, Pattern Recognition, Feature Discrimination

Sensitive Periods
Developmental windows of heightened receptivity.
Developmental Timing, Personalized Learning, Learning Readiness

Simple to Complex
Progressive learning from foundational to advanced.
Scaffolded Thinking, Complexity Navigation, Analytical Growth

Three Period Lesson
Informal formative assessment in three parts.
Formative Assessment, Recall Development, Language Precision

Timelines
Chronological visual tools for historical study.
Temporal Reasoning, Chronological Thinking, Visualization

Whole to Parts to Whole
Contextualizing detail within larger frameworks.
Holistic Analysis, Contextual Reasoning, Integrative Thinking

Montessori-Aligned Terms for a Digital and Emerging-Technology World

As the world becomes increasingly digital, Montessori education continues to offer a grounded and human-centered path forward. These new terms extend Montessori principles into emerging technological spaces without rushing the child or replacing hands-on, sensorial experience. They are not about teaching technology for its own sake. Instead, they help educators prepare a thoughtful, balanced, and ethical digital environment that protects concentration, supports human relationships, and honors the child's developmental needs.

Each term reflects a Montessori value applied to modern tools: awareness over automation, human agency over convenience, observation over reaction, and purposeful use over passive consumption. Technology, when introduced slowly and intentionally, becomes a material for exploration—not a distraction or substitute for real experience.

These concepts help Montessori guides navigate the digital world with clarity and confidence, ensuring the child remains the center: capable, grounded, imaginative, and free to build a meaningful relationship with both the natural and technological worlds.

Montessori Term | Definition | AI Competency Alignment (Extended)

Absorbent Mind
A formative stage where the child naturally takes in impressions from the world.
Curiosity, Cognitive Flexibility, Self-Directed Learning

Attention Ecology
Honoring attention as a precious developmental resource and protecting it from depletion.

Attention Management, Cognitive Agency, Healthy Tech Habits

Algorithm Craft
Creating simple logical sequences through hands-on and digital exploration.
Sequencing, Computational Thinking, Logic Design

Agent Awareness
Recognizing when technology acts autonomously and when human intervention guides it.
Autonomy Oversight, Safety, Human-in-the-Loop Thinking

Augmented Observation
Using digital overlays only to deepen real-world observation, not replace it.
Extended Perception, Data-Informed Awareness, Responsible AR Use

Curated Digital Tools
Selecting only purposeful, developmentally aligned digital materials.
Tool Literacy, Intentional Tech Use, Decision Quality

Data as Material
Gathering, classifying, representing, and interpreting information as a concrete work.
Data Literacy, Organization, Classification

Digital Grace & Courtesy
Practicing respect, empathy, consent, and kindness in digital interactions.
Digital Citizenship, Prosocial Behavior, Consent Culture

Digital Stewardship
Caring for digital spaces, data, devices, and shared technology responsibly.
Digital Responsibility, Security Awareness, Ethical Tech Practice

Embodied Computing
Understanding machines through physical interaction with sensors, motion, and cause-and-effect.
Robotics Foundations, Systems Thinking, Feedback Literacy

Embodied Tech Balance
Maintaining rich sensory, social, and physical experience alongside technology.
Well-Being, Embodied Cognition, Balanced Engagement

Ethical Automation
Choosing when technology assists and when human presence is essential.
Ethical Reasoning, Values Judgment, Human-Centered Computing

Guided Reality Expansion (AR/VR)
Introducing immersive tools slowly and sparingly to extend imagination while preserving inner vision and grounding.
Controlled Immersion, Spatial Reasoning, Context Awareness

Human-in-the-Loop Learning
Children remain thinkers, observers, and decision-makers alongside technology.
Human Judgment, Oversight, Co-Agency with AI

Material Intelligence
Exploring how intelligence emerges through interaction with real materials, movement, and environment.
Embodied AI Understanding, Systems Awareness, Constructive Curiosity

Mindful Digital Work Cycle
Engaging with technology purposefully, followed by reflection and return to the physical world.
Executive Function, Self-Regulation, Metacognitive Awareness

Practical Technology Life

Caring for, maintaining, repairing, and responsibly using tools — digital and physical.
Tool Fluency, Maintenance Skills, Technical Responsibility

Sensorial Coding
Making computation visible, audible, and tactile so logic is experienced through the senses.
Pattern Recognition, Sensory Integration, Conceptual Mapping

Thoughtful Immersion (VR)
Entering simulated environments with clear purpose, limited duration, and guided return to reality.
Reflection, Emotional Safety, Cognitive Anchoring

Transparent Machines
Understanding what a machine does, how it learns, and why it makes choices.
Explainability, Model Understanding, Bias Awareness

These terms offer a gentle Montessori vocabulary for navigating emerging technology while preserving the child's dignity, attention, and connection to the real world. They are a starting point, not a prescription. As research grows and our lives continue to evolve, Montessori educators and communities will naturally adapt, refine, and create language that reflects lived experience, developmental wisdom, and the needs of the children in their care.

Chapter 76

Appendix B: Glossary of AI & Education Terminology

As artificial intelligence becomes woven into the daily fabric of learning, new language emerges to describe how humans and machines think, act, and create together. This glossary serves as a bridge between the enduring human principles of Montessori education and the evolving vocabulary of AI. Each term has been carefully selected and adapted for educational relevance, aligning clarity with precision.

While this is not an exhaustive list, it offers a meaningful starting point for educators to begin the conversation, both for themselves and with their students, about what it means to live, learn, and act intelligently in an age shaped by algorithms.

Students aged 10–18 years should be encouraged to create their own set of vocabulary cards for AI education terminology, deepening understanding as they encounter and apply these terms in practical, hands-on contexts.

AI Education | Term | Definition | Basic

AI Ethics & Self-Reflection
The study of moral principles guiding artificial intelligence, ensuring responsible development and use, combined with the ability to evaluate personal decisions and actions in relation to AI's societal impact.

Algorithm
A step-by-step procedure for solving a problem or accomplishing some end, especially by a computer.

Algorithmic Thinking
The process of solving problems using a series of steps (algorithms) that can be implemented by a computer.

Analysis
A detailed examination of anything complex in order to understand its nature or to determine its essential features.

Application
The act of putting something to use; a use to which something is put.

Artificial Intelligence (AI)
The capability of a machine to imitate intelligent human behavior.

Attention
The act or state of applying the mind to something.

Awareness
The quality or state of being aware: knowledge and understanding that something is happening or exists.

Basic
Forming or relating to the most important part of something; fundamental.

Cause

A reason for an action or condition; something that brings about an effect or a result.

Citizenship
The quality of an individual's response to membership in a community.

Competency
The quality or state of having sufficient knowledge, judgment, skill, or strength (as for a particular duty or in a particular respect).

Computational Thinking
A problem-solving process that includes characteristics such as logically ordering and analyzing data and creating solutions using a series of ordered steps (algorithms).

Conceptual
Of, relating to, or consisting of concepts or ideas.

Coordination
The process of organizing people or groups so that they work together properly and well.

Creation
The act of making or producing something that did not exist before.

Data
Factual information (such as measurements or statistics) used as a basis for reasoning, discussion, or calculation.

Data Analysis
The process of examining, cleaning, transforming, and modeling data to discover useful information.

Data Structuring
The process of organizing data, typically in databases, to make it more useful.

Data Visualization
The graphical representation of information and data.

Development
The act, process, or result of developing (the process of growing or causing something to grow or become more advanced).

Digital
Involving or relating to the use of computer technology.

Digital Citizenship
The responsible use of technology by anyone who uses computers, the internet, and digital devices to engage with society on any level.

Effect
A change that results when something is done or happens; an event, condition, or state of affairs that is produced by a cause.

Empathy
The action of understanding, being aware of, being sensitive to, and vicariously experiencing the feelings, thoughts, and experiences of another.

Ethical
Involving or expressing moral approval or disapproval; conforming to accepted standards of conduct.

Foundations
A basis (such as a tenet, principle, or axiom) upon which something stands or is supported.

Hand-Eye Coordination
The way that one's hands and sight work together to be able to do things that require speed and accuracy.

Intelligence
The ability to learn or understand or to deal with new or trying

situations; reason; also: the skilled use of reason.

Learning
The activity or process of gaining knowledge or skill by studying, practicing, being taught, or experiencing something.

Logic
A proper or reasonable way of thinking about or understanding something; a particular way of reasoning or thinking about something.

Machine Learning
The process by which a computer is able to improve its own performance (as in analyzing image files) by continuously incorporating new data into an existing statistical model.

Mapping
The act or process of making a map; also: function 5a (a rule of correspondence between two sets that associates each element of the first with one or more elements of the second).

Materials
The elements, constituents, or substances of which something is composed or can be made.

Mathematical
Of, relating to, or involving mathematics.

Montessori
A child-centered educational method emphasizing independence, hands-on learning, self-paced exploration, and the prepared environment.

Motor Skills
The ability to perform complex muscle-and-nerve acts that produce movement; fine motor skills involve small movements like picking up small objects, while gross motor skills involve larger movements

like running.

Natural Language Processing (NLP)
A branch of artificial intelligence that helps computers understand, interpret, and manipulate human language.

Optimization
An act, process, or methodology of making something (such as a design, system, or decision) as fully perfect, functional, or effective as possible.

Order
The arrangement or disposition of people or things in relation to each other according to a particular sequence, pattern, or method.

Pattern
A repeated form or design; especially: one that is used to decorate something.

Precision
The quality of being precise; exactness or accuracy.

Problem-Solving
The process or act of finding a solution to a problem.

Processing
The act of taking something through an established and usually routine set of procedures to convert it from one form to another.

Project-Based AI Applications
A method of learning in which students apply artificial intelligence concepts to real-world projects, fostering critical thinking, problem-solving, and hands-on experimentation.

Project-Based Learning
A teaching method in which students learn by actively engaging in real-world and personally meaningful projects.

Recognition
The act of accepting that something is true or important or that it exists; the act of knowing and remembering upon seeing.

Reflection
Careful thought about something; an idea or opinion that is formed or a remark made as a result of careful thought.

Sequence
A continuous or connected series; the order in which things follow each other.

Simple
Easily understood, performed, or lacking complexity.

Skills
The ability to apply knowledge and expertise effectively in performing a task or activity.

Social
Related to interaction, communication, and relationships between individuals or groups.

Sorting
The process of arranging or categorizing objects, data, or concepts based on specific criteria.

Spatial
Relating to space, positioning, and the relationships between objects within an environment.

Steps
Individual movements or actions that contribute to progress in a process or journey.

Symmetry
The balanced and proportionate arrangement of elements, often

seen in design, mathematics, and nature.

Tactile
Related to the sense of touch or the perception of physical textures and surfaces.

Technology
The practical application of knowledge, especially in a particular area; a capability given by the practical application of knowledge.

Thinking
The process of using reasoning, imagination, and cognition to form ideas and solve problems.

Visualization
The formation of mental images or graphical representations to understand, analyze, or predict concepts and data.

Expanded Terminology for Computational Thinking

Computational thinking deserves its own place in this glossary because it reflects what Dr. Maria Montessori called the *mathematical mind,* the human tendency to seek order, sequence, and logic in the world (Montessori, 1949/1995). It is more than coding; it is a disciplined yet creative way of thinking that empowers learners to analyze, predict, and design. In Montessori education, these same cognitive patterns emerge when children classify, sort, build, and experiment with concrete materials. By breaking down complexity, recognizing patterns, abstracting essential ideas, and constructing clear sequences, students engage both intellect and imagination. When woven through the Great Lessons and applied in real projects, computational thinking becomes a natural extension of the child's cosmic task: to understand the systems of life, contribute intelligently, and act with responsibility in an interconnected, technological world.

Computational Thinking | Term | Definition | Basic

Abstraction
Focusing on the essential information needed to solve a problem while filtering out unnecessary details.

Algorithm Design
Creating a clear, step-by-step set of instructions to solve a problem or complete a task.

Automation
Using technology or machines to perform repetitive tasks with minimal human input.

Computational Thinking
A problem-solving process that includes logically ordering and analyzing data, and creating solutions using a series of ordered steps (algorithms).

Debugging
Finding and correcting errors in code, logic, or data to ensure accurate outcomes.

Decomposition
Breaking down complex problems or systems into smaller, more manageable parts.

Evaluation
Assessing whether a solution, algorithm, or system effectively meets its intended goals and identifying areas for improvement.

Generalization
Applying solutions or insights gained from one problem to new or broader situations.

Iteration
Repeating a process or set of steps to refine, test, and improve a result.

Pattern Recognition
Identifying similarities, trends, or regularities within data or problems to simplify solutions and make predictions.

Prompt
A question, instruction, or statement given to an AI model to guide its response or output.

Simulation
Using a model or program to imitate real-world systems or processes for testing, exploration, or prediction.

Computational Thinking Terminology (Ages 6–12) — Foundational AI Literacy

Understanding the language of artificial intelligence begins with a few essential ideas that reveal how machines learn, decide, and interact. These terms form the groundwork for digital awareness, helping children recognize that every intelligent system, from a simple chatbot to a complex neural network, follows human-designed rules and values. When students learn what an algorithm is, how bias can appear in data, and why ethics guide technology, they begin to see AI not as magic but as a set of human choices expressed through code. This awareness nurtures responsibility, curiosity, and discernment, the same qualities Montessori education seeks to cultivate in every area of learning (AI4K12, 2022; UNESCO, 2023).

Ages 6–12 |Term | Definition | Foundational AI Literacy

Bias
A tendency in data or algorithms that can lead to unfair, inaccurate, or unbalanced outcomes.

Chatbot
A computer program designed to simulate conversation with humans, often using natural language processing.

Dataset
A structured collection of data used to train or test an AI model.

Digital Footprint
The record of information, data, and actions left by one's online activity.

Ethics
The study of moral principles guiding decisions and actions; in AI, this involves fairness, accountability, transparency, and respect for human values.

Neural Network
A system of algorithms modeled after the human brain that helps computers recognize patterns and make predictions.

Computational Thinking Terminology (Ages 12–18) — Applied AI & Ethics

As students move into adolescence, their capacity for abstraction, moral reasoning, and independent thought deepens. Understanding how artificial intelligence works, and the values guiding its design, becomes essential preparation for responsible citizenship in a digital world. These terms help learners see how data, algorithms, and human ethics intertwine to shape decisions made by intelligent systems. Concepts such as machine learning, neural networks, and large language models reveal the architecture of modern AI, while ideas like transparency, bias, and ethical responsibility invite reflection on its societal impact. For Montessori education, this integration of technology and ethics aligns with the adolescent's developmental quest for purpose, justice, and contribution to the greater good (OECD, 2023; UNESCO, 2023).

Ages 12–18 | Term | Definition | Applied AI & Ethics

AI Agent
An autonomous or semi-autonomous system that can perceive its environment, make decisions, and take actions to achieve specific goals, often interacting with humans or other agents.

Algorithmic Bias
Systematic unfairness that arises when algorithms reflect or amplify human or data bias, leading to unequal outcomes.

Artificial General Intelligence (AGI)
A theoretical form of artificial intelligence capable of understanding, learning, and performing any intellectual task that a human can do.

Artificial Narrow Intelligence (ANI)
Artificial intelligence designed to perform specific tasks within a limited domain, such as translation, navigation, or image recognition.

Augmented Intelligence
A model of human–AI collaboration emphasizing that AI enhances, rather than replaces, human judgment, creativity, and decision-making.

Autonomous System
A system or machine that operates independently once activated, using sensors, data, and algorithms to make decisions without direct human control.

Computer Vision
A field of artificial intelligence that enables computers to interpret and process visual information from the world, such as images or video.

Data Privacy
The right and practice of controlling how personal data is collected, stored, shared, and used in digital systems.

Deep Learning
A subfield of machine learning that uses multilayered neural networks to analyze vast amounts of data and identify complex patterns.

Edge Computing
The practice of processing data closer to where it is generated (for example, on a local device) to improve speed, efficiency, and privacy.

Ethical AI
The practice of developing and using artificial intelligence responsibly, ensuring fairness, accountability, transparency, and respect for human dignity.

Explainability (XAI)
The degree to which the inner workings and decisions of an AI system can be understood, interpreted, and explained by humans.

Federated Learning
A method of training AI models across multiple devices or servers without sharing raw data, improving privacy and security.

Generative AI
AI systems that create new content, such as text, images, video, or music, by learning from existing data patterns.

Human-in-the-Loop (HITL)
A design approach where humans remain actively involved in AI decision-making processes to ensure ethical oversight and alignment with human values.

Inference
The process by which an AI system applies what it has learned from data to make predictions, classifications, or decisions.

Large Language Model (LLM)
A type of AI trained on extensive text datasets to understand, generate, and manipulate human-like language.

Machine Learning
A branch of AI that enables computers to learn from data and improve their performance over time without explicit programming.

Model Training
The process of teaching an AI system to recognize patterns or make predictions by exposing it to large amounts of data.

Multimodal AI
Artificial intelligence capable of processing and integrating different types of input, such as text, images, audio, and video, into a unified understanding.

Natural Language Processing (NLP)
A branch of AI focused on enabling computers to understand,

interpret, and generate human language.

Optimization
The process of improving an algorithm or system to achieve the most efficient, effective, or accurate outcome.

Predictive Analytics
The use of data, algorithms, and machine learning to forecast trends, behaviors, or future outcomes based on historical information.

Prompt Engineering
The practice of crafting and refining prompts to elicit specific, accurate, or creative responses from AI models.

Quantum Computing
An emerging field of computing that uses quantum mechanics to perform calculations much faster than traditional computers.

Reinforcement Learning
A type of machine learning in which an AI system learns through trial and feedback, receiving rewards or penalties for its actions.

Responsible AI
The development and deployment of AI systems that are lawful, ethical, sustainable, and aligned with human and environmental well-being.

Singularity
A speculative future point when artificial intelligence surpasses human intelligence, leading to unpredictable technological or societal transformation.

Supervised Learning
A type of machine learning where models are trained on labeled data to make predictions or classifications.

Synthetic Data
Artificially generated data created by algorithms to train or test AI models when real data is limited, sensitive, or unavailable.

Transparency
The principle that AI systems should be understandable and explainable to humans, especially regarding how data and decisions are handled.

Unsupervised Learning
A type of machine learning where AI models identify hidden patterns or structures in unlabeled data without explicit guidance.

Made in the USA
Las Vegas, NV
24 October 2026

56481315R00224